MARINE PLANKTON

MARINE PLANKTON

a practical guide

G. E. NEWELL
Formerly Department of Zoology, Queen Mary College, University of London

and

R. C. NEWELL
Department of Zoology, University of Capetown, South Africa

2006 - 2023 edition, published by:
Pisces Conservation Ltd.,
Office 3, The Granary, Newlands Manor Farm
Everton, Hampshire, SO41 0BG, UK
+44 (0)1590 674000
pisces@pisces-conservation.com
www.pisces-conservation.com

ISBN 978-1-904690-42-9

First published (Hutchinson & Co, London) 1963
Second edition (Hutchinson & Co) 1966
Third edition (Hutchinson & Co) 1967
Fourth edition (Hutchinson & Co) 1973
Fifth edition (Hutchinson & Co) 1977

Contents

Figures in Text

Plates

Preface

The purpose of this book is to give students of zoology a concise account of the kind of practical study of plankton they might make at sea or in the lab. It discusses methods of plankton collection, sorting and quantitative estimation. Sufficient information is given in the form of notes and diagrams for the commoner members of the British plankton to be identified, at least provisionally, without recourse to dichotomous keys which have been omitted partly to save space and cost.

It should be stressed that this is in no way a book for specialists. It is designed as a hand book for students, but for a book of this nature the bibliography is rather full-although it has been subjected to a rigorous selection. We hope that sufficient 'key works' are included for references to earlier, sometimes essential, publications to be traced by those who wish to pursue their study in depth. The bibliography is split into two sections: (1) gives references of general interest and is arranged in alphabetical order under authors' names; (2) deals with works mainly of taxonomic interest arranged under group headings. There is a full list of contents and the reader with a working knowledge of invertebrate zoology should readily find his way about the book without recourse to the index.

A few works of general interest deserve special mention. First there is Sir Alister Hardy's splendid book, *The Open Sea, the World of Plankton* (1956) which gives so much essential information so attractively that it should certainly be read at an early stage by every student. Also very valuable, although written from a different point of view, is James Fraser's *Nature Adrift* (1962). Trégouboff and Rose, *Manuel de Planctonologie Méditerranéenne* (1957) is a useful treatise, although it deals with Mediterranean plankton. Russell and Yonge's *The Seas* can be strongly recommended for a general background to the study of marine life. The *Fiches d'identification*, edited by Jesperson, Russell *et al.*, provide compact authoritative guides to the identification and distribution of zooplankton and give references to all the important taxonomic literature. Although not yet complete, these sheets are indispensable to all serious students of plankton. The several volumes of *Nordisches Plankton* are, perhaps, works of reference mainly for specialists, but these and certain volumes of the *Faune de France* are referred to in the bibliography on numerous occasions, and for certain groups they are the main sources of information.

The list of references given in the bibliographies at the end of this book, although by no means exhaustive, bears witness to our indebtedness for information to many authorities in many parts of the world. In particular we wish to acknowledge our reliance on published illustrations in making the simple line-drawings we have used as illustrations. For, although many of these have been made from, or checked by, observations on actual specimens, it would have been manifestly impossible to have done this for all of them. We are fully conscious of the fact that all our figures are diagrammatic and often fall short of the standard set by works of reference and we would urge that these should be consulted by all who wish to carry their study beyond an elementary stage or by those who can enjoy the beauty of the illustrations given, for example, in Russell's *Medusae of the British Isles*, in the numerous papers of Dr Lebour and in the recent monograph on *The Eggs and Planktonic Stages of British Marine Fishes* by Russell. Our debt to the numerous contributors to the *Fiches d'identification* will be obvious to all those who are familiar with them.

Dr J. H. Wickstead made many valuable criticisms of the typescript and we are grateful to him for saving us from many errors. For such as remain we must take full responsibility.

G. E. AND R. C. NEWELL

PREFACE TO THE FIFTH EDITION

The need to reprint this book has provided an opportunity to revise the taxonomic sections and to include more diagrams where these have become necessary. In general, more examples of warm-water and Atlantic species have been included so that identification of organisms from the west and south west of the British Isles may be made. Chapter 13 has been completely revised following the publication of Russell's monograph on the eggs and larvae of British Marine fishes.

R. C. NEWELL

I
Introduction

THE term plankton is applied to all those animals and plants which live freely in the water and which, because of their limited powers of locomotion, are more or less passively drifted by water currents. This is not to say that they are incapable of swimming. On the contrary, the arrow worms, many crustacea and fish larvae, can swim well (Hardy and Bainbridge, 1954). But partly because of their small size, which makes their absolute speed slow, and partly because much of their movement is in a vertical plane movements from place to place in a horizontal plane are very limited.

Practically every major group of animals, either as adults, as larvae, or as both, has its planktonic representatives or planktonts, as they are sometimes called. The taxonomic range is paralleled by a great variety in size, for creatures such as bacteria, minute flagellates and other protozoa up to the larger jellyfishes, often some feet across the bell, are included in any wide definition. It is sometimes useful to grade planktonic organisms into size groups which are in part related to the mesh sizes of nets in common use. Thus, organisms readily visible to the naked eye and down to a size which is retained on a coarse net with mesh aperture of 1mm are referred to as 'coarse net' or macroplankton. Organisms which are below 1mm in size, but are large enough to be caught by a net with a mesh aperture of 0.075mm, form the microplankton, whilst all smaller than this form the nanoplankton, which, because it is often studied after centrifuging water samples, is sometimes called 'centrifuge plankton'. With rare exceptions, planktonic plants are too small to be included in the macroplankton, but nanoplankton consists almost entirely of bacteria and autotrophic flagellates. The smallest flagellates, that is those below 5 μ, are sometimes categorized as ultraplankton, whilst very large animals, such as large jellyfishes, are spoken of as the megaloplankton. Such terms as these have their practical importance, but a classification in biological terms is more often used. Two main categories, the phytoplankton, which includes all the plants, and the zooplankton, all the animals, are recognized. It is also convenient to speak of the holoplankton-those organisms which are planktonic throughout the whole of their life-histories, and the meroplankton-the developmental stages (mainly eggs and larvae but including also polymorphic forms such as hydroid medusae and a few instances, e.g. monstrillid copepods, whose adults are pelagic but whose larvae are not) of organisms whose adults dwell on the sea-floor and form part of the benthos. After a sojourn in the plankton, whose length varies from species to species, the larvae metamorphose and settle on the sea-floor. For most animals this means that a pelagic life is irrevocably abandoned, but in recent years it has been discovered that the settled juveniles of mussels resorb the byssus threads that attached them to the thread-like substrata (hydroids etc.) for which they have shown a preference, become free-floating and are transported by currents to resettle in grooves or niches in hard surfaces. This process may be repeated several times and in each phase young mussels appear in plankton samples (Verwey, 1954, de Blok and Geelen, 1958). Rather similarly, juvenile and even adult lugworms leave the bottom on occasions, float and re-burrow in new situations (Newell, 1948 and 1949a). Post-metamorphic juveniles of many other animals, e.g. ophiuroids, are also common in certain samples-even in some above deep water far from land. Perhaps these and many others, usually regarded as accidentally occurring in plankton, having been 'whirled up' from the bottom, are undergoing a second truly pelagic phase. Many small bottom-dwelling crustacea swim for a while in shallow water over the intertidal zone when the tide floods, and contribute to a very characteristic type of inshore plankton described by Colman and Segrove (1955). In short, a distinction between planktonic and benthonic organisms is less clear-cut than might be supposed, but in practice the terms retain their usefulness. Beyond this point the classification of planktonts is on a taxonomic basis and is considered in Chapters 3 and 4.

II
Methods used in the Study of Plankton

Collecting

THE most widely used apparatus for collecting plankton is a plankton net. This, despite many minor variations in pattern, consists essentially of a cone of bolting silk (or equivalent material) mounted on a ring or hoop to which are attached three thin rope bridles spliced on to a smaller ring by means of which the net can be shackled to a towing rope or warp (fig. 1). Modern nets of all patterns differ from this simple plan by having the first part of the net, i.e. the part attached

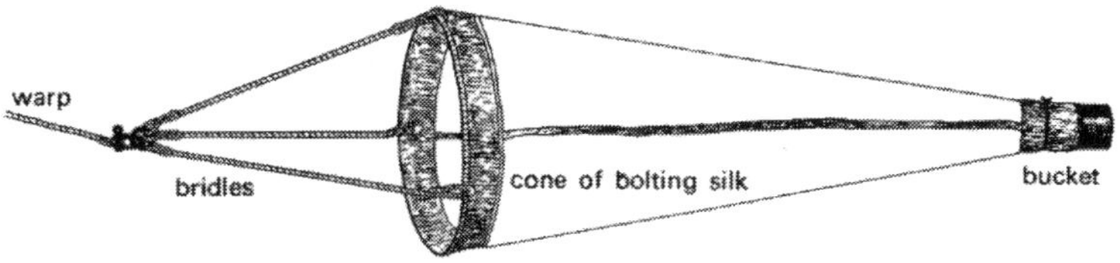

Figure 1. Standard plankton net

to the hoop, of thin canvas, thus making the net stronger. Also, the end of the cone is left open and is reinforced by strong material; tapes or cords are sewn to this so that a small metal or glass jar can be tied into it. The jar receives most of the plankton as the net is towed along, but some always remains on the wall of the net and is removed by turning the net inside-out and washing it in a wide-mouthed receiving jar, holding, say, about a litre of seawater. This sample is then concentrated after preserving fluid has been added to it.

There are several patterns of plankton net in common use today. Some of these are shown with explanatory notes in figs. 2, 3 and 4. They differ in their ability to catch macroplankton under standard conditions because of variations in the amount of water regurgitated through the mouth, for no net filters all the water which enters its mouth. In fact the so-called standard net is said to filter only one-tenth that filtered by the Hensen net, whilst the Heligoland net is four times as efficient as the Hensen net.

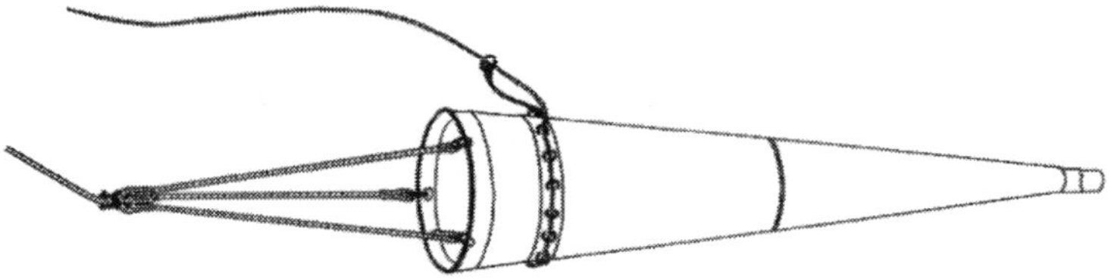

Figure 2. Net with closing device (towing)

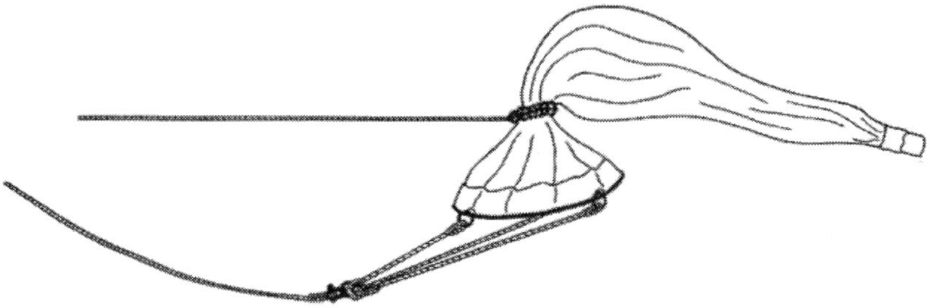

Figure 3. Net with closing device (being hauled in)

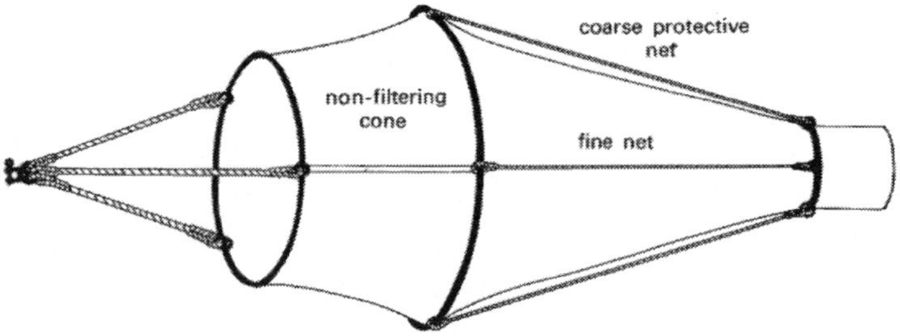

Figure 4. Hensen net.

Qualitative sampling of the surface waters is a simple matter. A suitable net is towed slowly (at about one to one and a half knots) behind a boat, and in British waters a ten-minute haul usually suffices to give an adequate amount of plankton. But the quantitative estimation of plankton bristles with difficulties. Ideally it would be hoped that the number of organisms in a standard volume of water-either a litre or a cubic metre is usually chosen-could be estimated, but this is never possible with any real accuracy. Firstly, as has been explained, some of the water gets spilled out before it has been filtered, so that the number of organisms caught is always less than that contained in the water entering the net. Particularly when working with a fine-mesh net in waters rich in phytoplankton, the meshes get progressively clogged and so the filtration rate falls off with time, more and more water spilling from the front of the net as resistance increases. If all the water entering the net was filtered then, of course, it would be possible to calculate the volume filtered by the simple formula $\pi r^2 l$ where r is the radius of the hoop at the front of the net and l the distance through which the net is hauled. Such a method of estimation can be, and is, used to give approximate results and is best suited for use with slow hauls made vertically through the water. Hauls of this type are often used to assess the amount of plankton in a given water column, but inaccuracies of other kinds creep

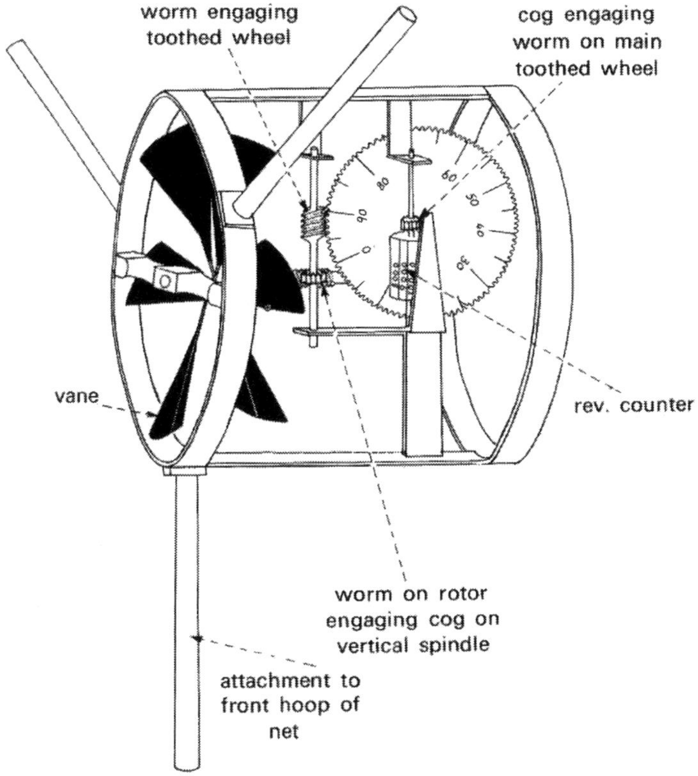

Figure 5. Harvey flowmeter

in here, for it is well-nigh impossible to keep a ship stationary at sea, so a truly vertical column cannot be sampled.

In order to overcome some of these difficulties some nets are fitted with a water-or flow-meter a little way behind the front. A well-known one is that designed by Harvey (1934) (fig. 5). It consists of a rotor driven by six thin brass vanes and a toothed wheel which engages in a worm drive on the spindle of the rotor. The toothed wheel has graduations around its circumference which enable the number of revolutions of the rotor to be recorded as the net is towed along, in much the same way as a cyclometer records the revolutions of a bicycle wheel. After it has been suitably calibrated the meter gives a measure of the water filtered, since it is this water which drives the rotor vanes. Widely used in many countries is the Clarke-Bumpus sampler (Clarke and Bumpus, 1950) which can now be obtained commercially. It is a small net with a mouth aperture 5in and a cone about 2ft long mounted at the rear of a brass tube of the same diameter which contains a flow-meter registering the amount of water filtered. A metal disc acts as a shutter across the entrance to the net when it is first lowered into the water but can be rotated open at any desired depth by a 'messenger' sent down the towing warp. The Clarke-Bumpus apparatus has been modified by various workers. Paquette, Scott and Sund (1961) improved its catching power by enlarging the net, whilst in 1957 Paquette and Frolander improved the opening-and-closing mechanism. These papers should be consulted for details of the newer patterns, although the original apparatus remains useful for some types of investigation. Small-mouthed nets, although used in most investigations, have the disadvantage that they fail to catch many of the larger and faster-moving planktonts such as fish larvae. Larger nets towed at higher speeds might seem an obvious way of getting over this difficulty, but, in practice, nets above a few feet in diameter are difficult to handle, whilst at speeds in excess of two or three knots the ring buckles. Isaacs and Kidd (1953) successfully met many of these objections by designing a mid-water trawl to which refinements have been made subsequently. Essentially this consists of a cone-shaped net bearing a V-shaped vane which acts to give negative pitch and so keeps the net below the surface. The most widely used models have bridle spreads of either 3 or 6ft (fig. 6), but larger ones are in use. It will be noticed that the use of meshes of three different sizes makes metering of the net impracticable.

Still in use is the Petersen young-fish trawl and its modification made by Clark (1940). These, as the name implies, are chiefly of use in sampling the macroplankton. Faced with the problem of obtaining quantitative samples of organisms too big to be coped with by the Clarke-Bumpus net, Currie and Foxton (1957) developed the Nansen net and fitted it with a flow-meter built on Harvey's principle. They also incorporated a modified Bourdon tube which acts as a pressure gauge and hence can be used to measure depth. By means of a stylus writing on a smoked-glass drum turned by the rotor, both the volume of water filtered and the depth were recorded

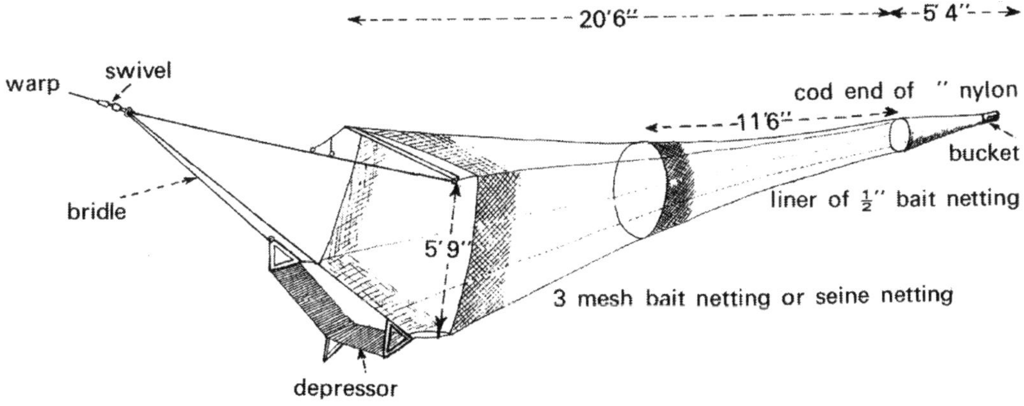

Figure 6. Isaacs-Kidd midwater trawl

as a helical trace in which the number of revolutions of the cylinder is proportional to the volume, and the vertical displacement of the trace gives a measure of the depth through which the net is hauled (fig. 7). Obviously the apparatus has to be calibrated before it is used and this is done either in a pressure tank or at sea or by both methods. The Currie-Foxton net, as has been implied, is used only for vertical hauls.

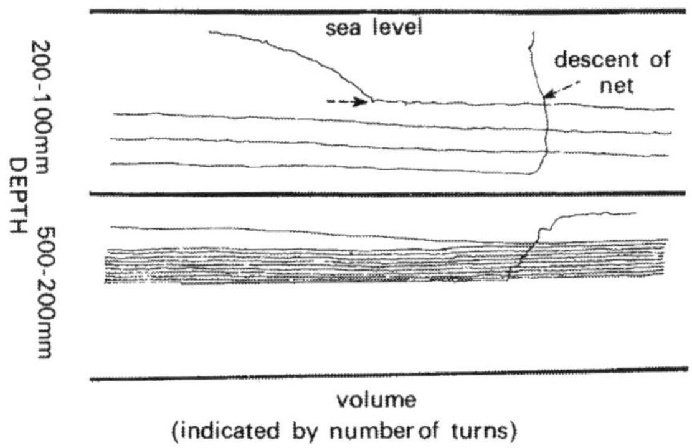

Figure 7. Flowmeter traces (after Currie and Foxton, 1957)

A full discussion of the problems involved in making quantitative plankton estimations by means of nets is given by Winsor and Clarke (1940), Aron (1961), Hansen and Anderson (1961) and by Yentsch *et al.* (1961). These papers should be consulted for a synoptic view and for the references given to other work in this field.

At first sight it might be thought that pumping known volumes of water through suitable filters would prove to be a satisfactory way of obtaining standard samples of plankton; but this also has its limitations, although it has often been used satisfactorily (see, for example, Thorson, 1946). Pump-sampling is ill-suited for catching fast-moving organisms such as euphausids and fish larvae which tend to avoid the entrance into the hose through which the water is pumped. Moreover, large organisms may get damaged when passing through the pump itself. In addition, a pump often misses organisms which occur in compact swarms, as some of them do. Lastly, a pump is obviously unsuitable for sampling plankton at depths greater than, say, 100m. Nevertheless, as pointed out by Aron (1958 and 1961), a plankton pump has 'several obvious advantages over other kinds of sampling equipment'. The amount of water filtered in obtaining each sample can be accurately determined and clogging can be virtually eliminated. Moreover, the depth at which samples are taken can be accurately determined, whilst at the same time water samples for hydrographic purposes, such as chemical analysis and temperature, can be obtained. It may well be that with improved pumps even the larger, fast-moving macroplankton could be accurately sampled, but at present it seems that pump-sampling is most useful for estimation of the microplankton only (see Banse, 1955, 1956, 1959; Barnes, 1949a).

Hardy (1936, 1939) has invented an apparatus called the continuous plankton recorder which can be towed behind ships at full speed (fig. 8). It is now in use on several shipping lanes and its special purpose is to record the distribution of plankton over large distances. Extremely valuable results, important to fisheries as well as of scientific interest, are being obtained with it and are published in the *Bulletins of Marine Ecology*. Earlier (1926), Hardy had made a simpler apparatus, the plankton indicator, for use by the skippers of herring drifters, enabling them to spot water rich in *Calanus*, one of the main foods of the herring, and hence to shoot their nets in areas likely to yield a heavier catch. Its value for this and for more strictly scientific purposes is described by Hardy (1956), by Glover (1953, 1961) and by Miller (1961). See also Barnes (1958).

Other pieces of apparatus which sample the plankton at high speed have been

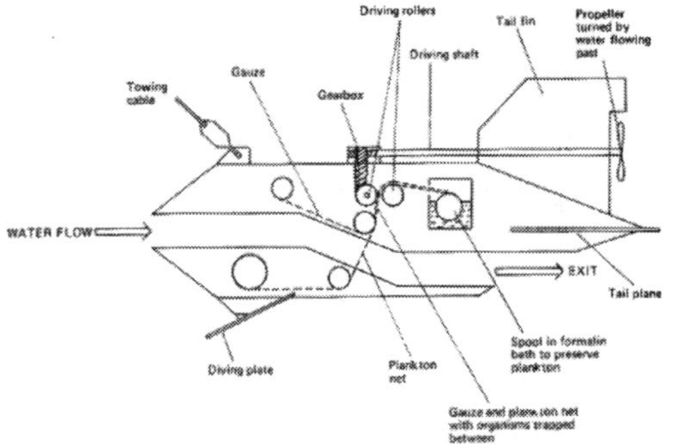

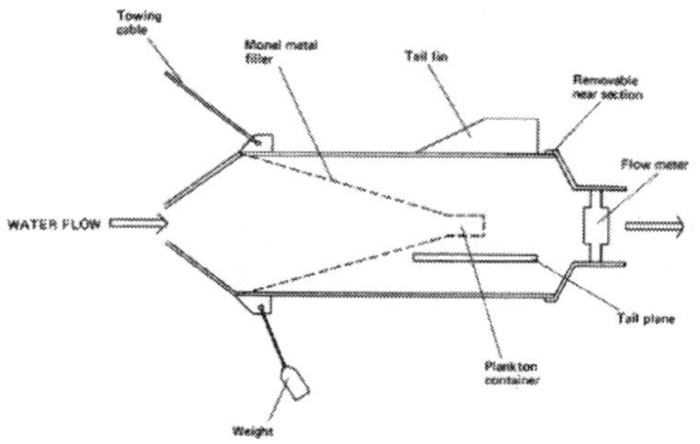

Figure 8. A Hardy continuous plankton recorder
Figure 9. A Gulf III high-speed plankton sampler

discussed by Gehringer (1961). One of these is the so-called 'Gulf III' made entirely of Monel metal (a cupro-nickel alloy) which resists corrosion, even the filtering device being made of a mesh of this metal (fig. 9). A modified Gulf III (called a T.T.N. by English workers, meaning 'tin tow-net') has been successfully used by workers from several European laboratories; Southward (1962), for example, in the Plymouth area. Descriptions of this and other high-speed samplers are given by Gehringer but are obviously outside the scope of this small book. For the record only, mention may be made of the self-propelled Research Vehicle which is being developed by the Applied Physics Laboratory of the University of Washington, designed primarily for obtaining continuous data on sea-floor topography but with instrumentation to record also oceanographic and geophysical data (Van Wagenen and O'Rourke, 1960). This vehicle has its course controlled by sonic pulses emitted from a ship, and its biological applications have been realized by fitting it with a Clarke-Bumpus plankton-sampler. In modified form it my well prove to be an important adjunct to sampling plankton in conjunction with the collection of hydrographic data. Its uses are discussed by Aron (1961).

Great interest attaches to the horizontal stratification of plankton, that is, to its qualitative and quantitative variation with depth, and such investigations have presented problems which have been solved, or partially solved, in a variety of ways. One is to make hauls at various depths and, after analysis, to subtract from the deep hauls all the organisms found in the shallower hauls. The limitations of this method are obvious. Clearly the best method would be one which closed the net at a known depth or, better still, opened it at a known depth and closed it again before it was hauled to the surface. A completely reliable method has not been devised. Instead, various devices which close the net at its mouth, some way down the net, or at the entrance to the bucket have been used. These methods have been fully reviewed by Currie (1961) and by Yentsch *et al.* (1961). As Currie points out, any mechanism of closure should ensure that there is no loss of the catch on closure and that there is no contamination from water through which the catch is towed either before or after closure. For smaller nets some form of closure in the form of valves (as in the Clarke-Bumpus net) works well, but for larger nets used for active macroplankton the mechanical forces which have to be overcome are so large as to render flap-valves impracticable in nets over 50cm in diameter. Yet, since sampling of oceanic plankton usually requires the use of large diameter nets, some sort of 'throttling device' is essential. The Nansen method relies on throttling the net at a little distance behind the mouth by disengaging the towing bridles and towing the net from 'a rope which encircles the net in such a manner that it tightens round the net when strain is applied' (fig. 2). The limitations of this method have been discussed by Barnes (1949b). Its chief danger lies in loss of catch after the net has begun to close, some water flowing out of the mouth because of the positive pressure inside the net when it is fishing. To get over this difficulty Currie and Foxton (1957) designed their net so that the effective filtering part of the net lay well behind the point of closure, whilst the net itself, having a good filtering coefficient, did not develop much internal pressure. Bucket-closing systems are in the experimental stage. If perfected it might become possible to use several buckets on one net, and if suitable opening and closing mechanisms were developed a succession of samples from different depths could be taken with one lowering of the net. This would save a great deal of time. The chief difficulty lies in assessing accurately how much of the plankton actually gets into the bucket and how much remains on the wall of the net. An alternative to having several closeable buckets for one net is to use several nets, each of which can be closed by one type of device or another. Commonly the net is closed by a 'messenger' sent down the towing warp, as in the Nansen net modified for use by the Discovery Expedition. This is a practically fool-proof device for vertical hauls, but less applicable for the oblique or horizontal ones sometimes required to catch the macroplankton when greater volumes of water must be filtered to give quantitative results. For this purpose, perhaps, the use of multiple nets, opened by devices which respond to pressure - itself directly related to depth - would seem to offer the best lines for future development. For the sampling of layers less widely separated in depth, electrically or sonically controlled opening-and-closing devices may be successfully developed in the future.

All the methods mentioned above give results which are valuable for making comparisons between the plankton of different waters, but none is strictly quantitative, since all the catches are minimal. As Harvey (1934) points out, 'the only strictly quantitative method of sampling the population is to centrifuge a small sample of water and count the plants in it'. This method and methods of culturing are, however, laborious and quite unsuitable for dealing with zooplankton, so they are used only by investigators of very small planktonic organisms - most of which will pass through even the finest bolting silk. Special methods for bacteria are given by Zobell (1941, 1946).

It is at once obvious that the mesh-size of the material of which a net is constructed will influence the kind of plankton caught. The coarser the net, the less the impedence to filtering, and so, since maximum filtration is desirable, the largest mesh-size consistent with catching particular organisms should be chosen. For many purposes it is usual to tow two or even three nets, made of bolting silk of different grades, simultaneously on the same warp, the choice of grades depending on the investigator's judgement. In practice, the macroplankton is usually sampled by a wider-mouthed net-the Petersen young-fish trawl, made of 1mm mesh stramin net. Otherwise, as stated, the net was until recently made of bolting silk, a material which is used commercially for sieving powders of various kinds. It is supplied[1] in twenty grades (see table), the coarsest of which has a mesh aperture of 1.250mm with eighteen meshes per inch and is numbered 0000 in the series. The finest is number 25. with a mesh aperture of 0.0535mm and 200 meshes per inch (actually 197).

[1] By John Stanier & Co., Silk Merchants, Manchester Wire Works, Manchester.

Table of mesh apertures and number of threads per inch of bolting silk

No.	*Threads per inch*	*Average dimension of aperture in inches*	*Average dimension of aperture in millimeters*
0000	18	0.0492	1.250
000	23	0 0368	0.935
00	29	0.0277	0.705
0	38	0.0202	0.515
1	48	0.0154	0.392
2	54	0.0134	0.342
3	58	0.0127	0.324
4	62	0.0120	0.306
5	66	0.0115	0.293
6	74	0.0094	0.241
7	82	0.0082	0.210
8	86	0.0075	0.193
9	97	0.0068	0.174
10	109	0.0059	0.150
11	116	0.0055	0.141
12	125	0.0053	0.135
13	129	0.0046	0.118
14	140	0.0039	0.100
15	149	0.00364	0.0925
16	157	0.00340	0.0865
17	163	0.0030	0.077
18	166	0.00293	0.0745
19	169	0.00281	0.0715
20	175	0.00273	0.0695
21	185	0.0024	0.063
25	197	0.0021	0.0535

Only this last grade will retain the bulk of the smaller diatoms. Few of the intermediate grades are in general use for plankton work. Commonly a coarse net (mesh aperture 0.324 mm), the international coarse-silk net or I.C.S.N. as it is usually called, a medium net (mesh aperture 0.0925mm) and a fine net (mesh aperture 0.063mm), suffice. A glance at a piece of bolting silk under a microscope shows how the warp and weft are specially woven so as to keep the mesh-size constant (fig. 10), and will serve to explain why this material is expensive. Bolting silk of much finer grades - down to 1000 meshes per inch - may become available in the near future and will provide a means of sampling all but the most minute plankton. Nylon

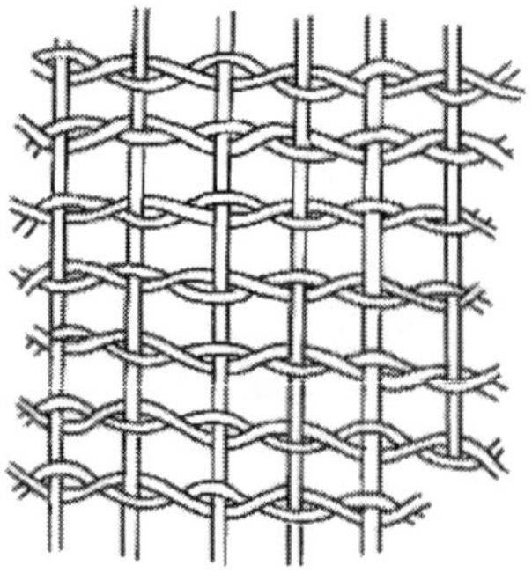

Figure 10. Bolting silk (seen under a microscope)

netting can also be obtained[1] and is particularly suitable for use in the tropics, where silk nets rot much quicker than elsewhere. Indeed, even in this country nylon plankton nets are superseding those made from silk. Supplied in the usual ranges of mesh-sizes up to 600 per inch (which retains even most protozoa) they are more robust and durable.

Preservation and storing of the catch

Although it is most instructive to examine the plankton alive, this is rarely possible, for unless it is kept cool (for example in a vacuum-flask) it soon dies. More usually, therefore, examination is left until after the sample has been preserved. Almost invariably the plankton should be fixed by the addition of neutral formaldehyde (Armstrong and Wickstead, 1962) or, alternatively of hexamine formalin, sufficient to bring the concentration to about 4%. This fluid suffices to preserve the sample indefinitely, and also has the effect of sending all the plankton to the bottom of the jar. As has been mentioned, a fairly large jar is needed to receive the plankton after the nets have been flushed - too large a jar to be stored conveniently. So, after the addition of formalin, the plankton is concentrated by filtering off a lot of the fluid through a fine-meshed net. It is advisable, however, not to concentrate the sample too much, but to leave it mixed with, say, about 200ml of fluid. The concentrated sample is then stored in suitable bottles or screw-topped jars, preferably plastic ones. The date, place of origin, mesh-size of the net, length and depth of the haul and any other relevant information should be written in Indian ink on good-quality paper and placed in the jar (labels on the outside usually peel off after a time). Information pencilled on a plastic screw-topped jar can save time when sorting, since a dense plankton sample obscures the label inside.

Counting the organisms

With the reservations mentioned previously, the concentrated sample represents the total plankton in the volume estimated to have been filtered by the net. For reasons which will appear later it is essential that the volume of the plankton plus the preserving fluid should be determined. This is usually done by tipping the contents

[1] From Henry Simon Ltd, Cheadle Heath, Stockport.

of the jar into a measuring cylinder, and if the mixture is allowed to stand for a day or so until the plankton has thoroughly settled the volume of the plankton itself can be measured at the same time. This gives a useful, even if rather rough, measure of the richness of the sample, but one which is easily falsified if a few really large planktonts, like jellyfishes, get included. These are usually taken out before measuring the volume. Alternatively, and more accurately, the volume of the plankton can be determined by the displacement method. First the total volume of the concentrated sample plus the preserving fluid is measured. Then the plankton is filtered off, using a filter paper in a funnel, and the volume of the filtrate is measured. The volume of the plankton is then obtained by the difference between the two volumes. A measure of the total catch can also be made by weighing the filtered plankton, a method which is preferred in many recent investigations. An index of the density of the plankton can be made by weighing an aliquot portion of preserved plankton after it has been filtered through filter paper, but a more acceptable method is to divide the fresh sample (i.e. before formalin has been added) into two equal portions. One is subsequently preserved for sorting and counting. The other is filtered, washed and dried at 50°C in a desiccator inside an oven and then weighed as rapidly as possible. The object of dividing the fresh sample into two halves is that drying ruins the plankton for subsequent examination. Fresh plankton can be stored for some time in a refrigerator or vacuum-flask cooled with ice, but the sample must not freeze.

Larger organisms, such as are rated as macroplankton, are usually present in numbers small enough for all of them to be counted, often merely by tipping the sample into a flat, shallow dish and examining it with the naked eye or with a low-power lens. This count should always be made first and the sample returned to the jar. Members of the microplankton, on the other hand, are nearly always far too abundant for total counts of them to be made. Instead, from the thoroughly mixed sample small aliquot portions are taken for examination and counting, after which, by a little simple arithmetic, the results can be expressed as numbers of organisms per litre or per cubic metre. This is easy in principle but more difficult in practice. For one thing the fixed organisms quickly sink to the bottom after the sample has been stirred; for another, the organisms vary greatly in size and what might be a convenient aliquot or sub-sample for one size range would be inadequate for another. A word about mixing must now be given! The total sample should be agitated - shaking will often suffice - but on no account should it be stirred or swirled in a way which will produce definite currents in the liquid, since these will have a sorting effect which will ruin the validity of sub-sampling, as can easily be demonstrated. All the smaller organisms tend to fly out to the periphery of a rotating fluid, whilst the larger ones tend to remain near the centre. A well-known device for withdrawing small volumes quickly from the well-stirred sample is the Stempel pipette (fig. 11), the usual pattern of which delivers 1ml at a time. It consists of a wide-mouthed syringe, the lower part of the piston shaped like an hour-glass, so that when the

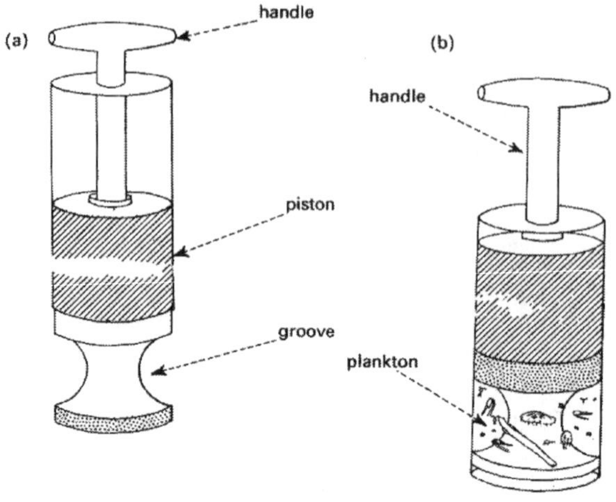

Figure 11. The Stempel pipette

piston is raised a known volume is trapped between it and the walls of the cylinder. This can then be squirted into any suitable dish, say a petri dish, for examination under a microscope. To facilitate counting, the dish can be placed over a sheet of graph paper which enables the material to be systematically worked over, a square at a time. Stempel pipettes are in less general use than formerly except for estimations of the smaller microplankton.

The counting tray originally made by Bogorov, and modified by Russell and Colman (1931), is an extremely useful device for sorting and counting plankton organisms. It consists of a plate of glass (black glass is often preferred) 6 in long and 3 in wide to which strips of bevelled glass are cemented by means of Canada Balsam to form three grooves, the outer two grooves being connected to the centre one at opposite ends (fig. 12). On the underside of the glass plate are cemented two thin glass rods which act as rails and prevent the plate sticking, even when wet, to the stage of a binocular microscope. A similar pattern of tray, but on a smaller scale,

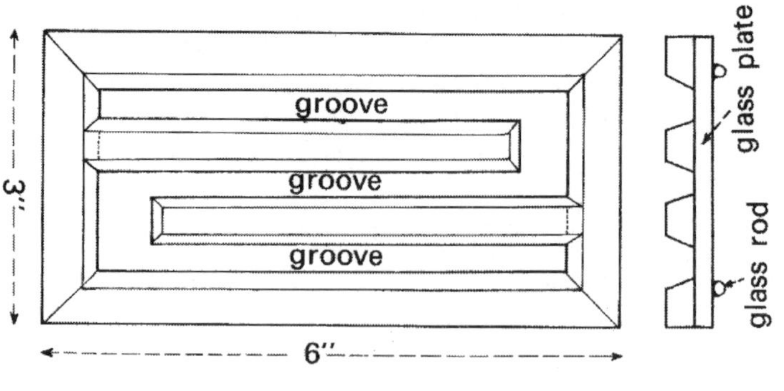

Figure 12. Counting tray

can easily be made by cementing narrow strips of glass cut from another microscope slide on to a large (3in x 2in) slide; Araldite is recommended as a cement.

Another simple device used in sub-sampling is the rafter cell (fig. 13). There are various patterns for this but all consist essentially of a rectangular frame of thin brass floored by a sheet of glass (a large microscope slide can be used), with a graticule of 1mm squares drawn on it with a diamond. The thickness of the brass frame and the dimensions of the glass are such that the cell has a volume of 1, 2 or even 3ml, whichever is the most convenient for the job in hand. For example, a 1ml cell is made from brass 1mm thick cut to enclose a rectangle 5cm long and 2cm wide. The cell is quickly filled with some of the well-mixed sample and immediately examined with a lens or a microscope. Indeed, with practice many of the larger members of the microplankton even can be recognized with quite low magnifications such as are given by simple binocular dissecting microscopes, but for many of the smaller

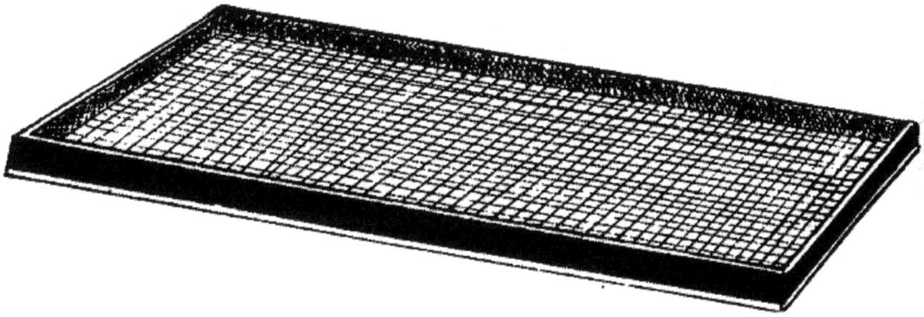

Figure 13. A rafter cell

organisms an ordinary compound microscope is required. It is often necessary to take out organisms one at a time from the dish or counting cell to look at them under high power and this is best done by means of a glass pipette fitted with a rubber teat. Consequently it is always as well to have a few of these with a variety of different-sized nozzles handy. Some workers find that loops of varying sizes made from thin tungsten wire mounted in needle-holders are more satisfactory for picking out organisms from a sample.

Another method which has recently been used successfully is to put the total catch into a wide, deep dish and dilute it to a fixed volume, say 5 l. The mixture is then agitated (*not* swirled) until it is thought that the plankton is uniformly distributed. This is then sub-sampled by ladles holding 10, 100 or 500ml, whichever is thought suitable, bearing in mind the abundance of the organisms. Only considerations based upon the statistical significance of the sub-samples can really determine which volume is appropriate. The ladled sub-sample is then filtered through bolting silk (it is advisable to use a finer grade than that used in the collection of the sample) to concentrate it into a volume suitable for estimating in a counting cell of one sort or another.

A refinement of this method is described by Wiborg (1951). This makes use of a special dish to receive the total catch before it is diluted. The dish is divided into a number of radial compartments and from each a tap can drain the sub-sample so provided into a filter for concentration and subsequent counting. Details of this method are given by Wiborg (1961). It is best suited to investigations of macroplankton but should, perhaps, be recommended only to the specialist.

Whatever method of sub-sampling and examination is adopted, the counting of individual organisms is bound to be a tedious business. It can be made easier and less liable to errors if *pro formas* are prepared which give a list down one side of the page of all the species likely to occur in the sample. Then as each organism is identified all that is needed to keep score is to tick in the appropriate column and to count the ticks when the sample has been worked over. Hensen, the father of quantitative planktonology, used a box divided into small compartments, one for each of the species he was counting. As he came across an organism he put a small lead shot or coin in the particular compartment. Nowadays inexpensive mechanical counters, which record numbers whenever a lever is pressed, can be bought and are useful when looking at one kind of organism; and for counting one species at a time.

To return to the estimation of numbers of organisms. If the volume of the concentrated sample was found to be 200ml, then, if the sub-sample was 1ml, obviously the number of organisms in the total catch is 200 times that counted in the sub-sample; and if the volume of water filtered by the net had been estimated to be 100m^3, then the number per m^3 is one-hundredth of this. Slightly more complicated calculations, but along the same lines, have to be made if abundant small organisms are to be counted. 1ml of the concentrated sample may contain so many of them that it would be very difficult to count them. Then it is advisable to dilute the 1ml. sub-sample to 100ml and to take 1ml from this diluted sub-sample for counting under the microscope. Many useful variations of these methods will suggest themselves. Obviously the greater the number of sub-samples examined, the more accurate will be the estimation of the numbers of organisms in the total catch. On the other hand, this may involve so much extra labour that the investigator will not be able to keep up with samples collected at regular intervals. It has been shown that, provided a suitable method of sub-sampling is adopted, a count on a single one is statistically acceptable. Methods of standardization of sampling zooplankton at sea have been reviewed in an important paper by Fraser (1966).[1]

Planning the taking and examination of plankton samples

In a thorough investigation of the plankton of any area much thought should be given at all stages to the best methods of sampling, counting and expressing the results. No one method, or any single type of sampling gear, will suit all needs. A good deal depends on experience and the resources available, but to produce significant results with economy of effort it is essential always to keep to the front of one's mind the object of the investigation. Is it, for example, to determine the productivity of a certain area? If so, then a quantitative assessment of the phytoplankton

[1]This work is to be reported in more detail in a U.N.E.S.C.O. monograph.

is of prime interest, and the plant nutrients and solar energy available are ancillary considerations. This is a highly specialized problem and cannot be dealt with here [for general accounts see Barnes (1958), Clarke (1946), Harvey (1942), Steeman Nielsen and Jensen (1957). The relevant literature is reviewed by Steele (1959)]. Plankton research is many-sided, but most aspects, whether purely scientific or related to fisheries, pollution of inshore waters or hydrography, require the accurate identification of planktonic organisms and often a quantitative assessment of them. But so many species may be present in any one sample that considerable time is required to identify and count them. For many purposes this is not necessary. It will sometimes suffice to lump together, say, all the diatoms, all the medusae, all the calanoid copepods and so on. For other purposes counts of individual species will be essential. There are several good general accounts of the plankton and Sir Alister Hardy's book (1956), with its excellent bibliography, is a delightful introduction.

Figures and plates

The sections which follow are an attempt to give some guidance in the identification of the commoner members of the British marine plankton. If they are representative of a major group then a few rare examples are given. All the diagrams are original in the sense that they have been drawn for this book and where possible they have been checked against actual specimens. But we have in the main built them up from information, either pictorial or written, given in standard works of reference and *it cannot be too strongly urged that wherever possible such books should be consulted before a conclusive identification is made.* Even then some doubt must remain and the final authority must be a competent systematist, since for some groups, notably copepods, a general correspondence with a figure or description is quite inadequate. Figures in the taxonomic section are referred to by two numbers. The first refers to the plate; the second to the figure itself. All the figures are diagrammatic and it must always be remembered that the appearance of many organisms changes considerably in the process of preservation. This is perhaps particularly true of medusae whose tentacles contract and whose bell shrinks and becomes distorted. Moreover, the structure changes greatly with the age of the organism. Obviously this is true also of size, and the scales given are intended merely as a rough guide.

It must be borne in mind that systematic plankton investigations have not been made all round our coasts, so that the information on the distribution of planktonic organisms must be regarded as tentative. Here, as in the geographical distribution of shore-dwelling organisms, the reported distribution is partly, at any rate, a function of the number of competent observers. In general the distribution of a particular organism must be expected to be wider than reported in the literature. For the sake of brevity much of the information in the following account is given in note form.

III
Phytoplankton

Introduction

THE phytoplankton includes all the plant-like organisms; plant-like in the sense that they are autotrophic and contribute directly to the food available in the surface waters by building up their protoplasm and food reserves directly from carbon dioxide and salts in solution in the sea. Of prime importance are phosphates and nitrates, although it has been shown that many accessory food substances are required and that some plants can utilize directly more complex substances, such as ammonia, as a source of nitrogen. The smallest autotrophic organisms, particularly the minute or 'μ-flagellates', are now thought to be of great importance as a source of food for many microphagous organisms, but these small flagellates are rarely seen in plankton samples collected by nets or pumps. Instead they can be examined and counted only by special methods such as those used by Dr Parke and her colleagues at Plymouth. These are outside the scope of this book. Indeed, taxonomically the smaller flagellates (**8**.3 and 4) are practically an unknown group of the systematic position of even some of the larger ones which do appear in ordinary samples is in dispute as a result of recent studies [see, for example, Parke (1949, 1961)].

Most obvious in size and numbers in samples of British plankton are the diatoms. They are particularly conspicuous in early spring (March to April) and in autumn (early September–early October), although in shallow, turbulent sea areas, in which a thermocline is never established during the warmer months of the year, some diatoms may be abundant at all seasons. Next in importance to the diatoms are the dinoflagellates. Other planktonic algae of lesser importance are *Halosphaera* (**7**.4), a unicellular yellow-green alga, the Silicoflagellates (**8**.3) and *Phaeocystis* (**7**.3), a brownish flagellate which secretes a slimy covering and often occurs in vast numbers. Coccoliths (**7**.5, **8**.5) are small flagellated planktonic stages, with a calcareous capsule, which also have a fixed filamentous stage attached to the substratum. These, which may on occasions play an important part in the economy of the sea, are also of interest in that they are the organisms primarily responsible for the laying down of the chalk in Upper Cretaceous times. In this brief account the terms 'alga' and 'flagellate' have been used loosely. Often, as has been said, the systematic position of the organisms is in some doubt, but it is now becoming customary to place *Halosphaera*, *Phaeocystis* and coccolithophorids in a group, Chrysophyceae, along with other flagellates such as *Chromulina*, *Isochrysis* etc.

Class BACILLARIOPHYCEAE - diatoms

Although showing a great diversity of size and form, diatoms are usually immediately recognizable by their cases or frustules of silica, often elaborately ornamented, and by their yellowish or brownish chloroplasts. There is a central mass of cytoplasm, containing the nucleus, joined by cytoplasmic strands to the wall of the cell. A feature unique to diatoms is that the frustule is in two parts or valves, one fitting snugly into the other rather like a box into its lid. Diatoms range in size upwards from a few μ to about 1mm but many of them can unite to form long chains, often several centimetres in length.

Two main sub-classes are recognized, Centricae and Pennatae, the distinction between them being drawn by a difference in the nature of the sculpturing on the frustule. In the Centricae the sculpture patterns radiate out from a central or from a lateral point or points. In the Pennatae the sculpture lines are in more or less straight lines - somewhat feather-like, in fact - hence the name 'pennate'. Moreover, many pennate diatoms, all those which can move independently, have a narrow slit, the raphe, running along one or both valves. In practice most plankton samples will contain only centric diatoms or at best only a few pennate ones which are for the most part dwellers on the bottom where many of them can move about in a peculiar jerky fashion. Plankton samples from shallow, turbulent water will, however, often contain benthonic diatoms which have been whirled up into the water. Moreover, some diatoms, sometimes called tychopelagic forms, make chains on the sea-floor which often break off and release clusters of cells that can be pelagic for long periods. Such a one is *Melosira*. Most species of diatoms are very widely distributed and few can be said to be typical of particular water-masses.

Sub-class CENTRICAE

Notes relating to some of the commoner centric diatoms are given on the pages which follow.

Melosira

Cylindrical cells each containing numerous chloroplasts. Frustules with a sculpturing of fine dots. The cells unite to form chains like a string of beads. Ptychopelagic and coastal. *M. monilifornas* (= *borreri*) typical of estuaries from northern N. Sea to W. Channel and Atlantic. Common (**1**.2).

Paralia

Short, disc-shaped cells with curved edges united to form straight chains. Ornamentation of roughly hexagonal markings. Numerous small chloroplasts. One species, *P. sulcata*, a tychopelagic diatom, widely distributed in the Atlantic and N. Sea to W. Channel (**1**.3). Prefers full salinity water.

Coscinodiscus

Simple, disc-like cells with the valves sculptured with hexagonal or pit-like markings. Numerous chloroplasts. There are many species, some of which are benthonic, some tychopelagic and others truly planktonic. Many are quite large, up to a few hundred in diameter. A few of those common in British waters are illustrated in **1**.1, 5, 7, 8.

Thalassiosira

The cells are somewhat like those of *Coscinodiscus* but provided with numerous fine spines, often larger than the cell. A truly planktonic genus, although characteristic of inshore waters.

T. decipiens has disc-shaped cells with curved spines; it unites to form chains with large spaces between the cells, which are from 12-40μ in diameter. Atlantic, N. Sea and Channel. Not very common (**1**.9). Mainly a northern species.

T. gravida is similar to the previous species but with shorter spines and somewhat larger when full size. Common in the N. Sea, Channel and Irish Sea (**1**.10). Mainly a northern species.

Lauderia

Short cylindrical cells united in straight chains with very little space between the cells. Sometimes spiny. Numerous chloroplasts. *L. borealis* and *L. glacialis* occur in the Channel, N. Sea and Atlantic, but are not common (**2**.1). Both are mainly northern species.

Skeletonema

The cells are lens-shaped with parallel spines around the margins which serve to unite the cells into straight chains. Only one or two chloroplasts in each cell.

S. costatum is a common species in British waters, particularly near the coasts. It is widely distributed, euryhaline and abundant in estuaries, particularly in the early spring. The small cells, only 7-15μ in diameter, join up to form long chains (**2**.2).

Leptocylindrus

Long cylindrical cells united to form straight chains. Chloroplasts few in number. *L. danicus*, with cells 6-12μ in diameter, occurs throughout the N. Sea and Channel (**2**.3).

L. minimus is smaller than the previous species. It seems to be an inshore species and occurs in the Channel, N. Sea and Atlantic.

Guinardia

Cylindrical cells slightly longer than broad with flat ends except for one notch. Numerous chloroplasts. Cells quite large (about 40-90μ in diameter) united in straight chains. *G. flaccida* is common in coastal water of the Atlantic, N. Sea and Channel (**2**.4).

Corethron

Cells with rounded ends and numerous spines some of which serve to join up the cells. Many small chloroplasts.

C. criophilum is typical of offshore waters but does occur in the Channel and N. Sea as well as in the Atlantic (**2**.5).

Rhizosolenia

Long, parallel-sided cells with an asymmetrical point at each end which may be very long or else quite short and rather blunt. Numerous chloroplasts. The cells can unite into chains. There are numerous species, many of which are common in British waters. A few are illustrated in **2**.6-10. Some of the larger ones, like *R. styliformis* (**2**.6), common in the N. Sea, may be repellent to plankton-feeding fishes.

Chaetoceros

The cells are usually oval in cross-section and have nearly flat ends. A pair of long thin spines is found at each end of the frustule, and these, by fusion with those of neighbouring cells, unite them into chains. Chloroplasts are variable in size and number. This large and variable genus is truly planktonic but some species are more typical of inshore waters than others. Many species occur in British waters.

C. densum is one of the commonest and occurs in N. Sea, Channel and Atlantic (**3**.1).

C. danicum, with a similar distribution, is, however, more plentiful in waters of lowered salinity (**3**.3).

C. decipiens flourishes mainly in cooler waters but is found all round our coasts (**3**.2).

C. teres again is a widely distributed, but mainly northern, species (**3**.4).

Biddulphia

The cells often occur singly or in twos and threes, although longer chains can be formed. The ends of the cells bear one or sometimes two pairs of projections. Valves often triangular in end view. Chloroplasts numerous. By far the most abundant diatom in shallow inshore waters all around our coasts. Probably several species are tychopelagic, although also able to flourish and reproduce in the plankton.

B. aurita is common in all parts of the N. Sea and Channel, especially in the spring (**4**.1).

B. mobiliensis (**4**.2) is common all round our coasts, but flourishes farther north than either *B. aurita* or *B. sinensis*.

B. regia is very similar in form to *B. mobiliensis* and is regarded by some as a large variety of it. Common all round our coasts but particularly in the southern N. Sea and estuaries entering into it (**4**.3).

B. sinensis is tolerant of wide changes of temperature and salinity. It is believed to be a fairly recent introduction to the N. Sea and has spread to our coasts since 1903 (**4**.4).

Other species widely distributed throughout the British Isles are shown in **4**.5-6.

Bellerochea

The broad, flat cells are united in chains by their corners. Chloroplasts are numerous. This is regarded as a tropical or sub-tropical diatom, but *B. malleus* (**5**.1) is often quite common in the Channel and probably in most parts of the N. Sea.

Cerataulina

B. pelagica (**5**.2) is an inshore diatom more characteristic of our southern waters but also common in some Scottish lochs.

Lithodesmium

The valves in end view are triangular but with wavy margins.

L. undulatum is a coastal species widely distributed but perhaps commoner in the south (**5**.3).

Ditylum

The cells are elongated and at each end there is a pair of (or else numerous small) spines and a long central one which serves to join up the cells. There are numerous chloroplasts. Cells 25-60μ in diameter.

B. brightwelli is common in the N. Sea, Channel and Atlantic, particularly in warmer coastal waters (**5**.4).

Eucampia

Cells, concave at each end, unite to form very characteristic spiral chains. Numerous chloroplasts.

E. zoodiacus, with cells 25-75μ in diameter, is common around our coasts but particularly in warmer water (**5**.5).

Streptotheca

Square, flat cells with a twist about the middle. Probably widely distributed but mainly in inshore waters.

S. thamensis has cells 40-130μ broad and is common in the N. Sea and Channel (**6**.1).

Sub-class PENNATAE

Fragilaria

Members of this genus form ribbon-like chains and are common in shallow water. There are one or two chloroplasts. The cells are very flat and the valves are ornamented with dots arranged in rows at right angles to the upper and lower surface (**6**.2).

Asterionella

The elongated cells are thicker at one end. They stick together by the thicker ends to form colonies of variable shape.

A. japonica forms star-shaped clusters which in turn unite into spirals (**6**.5).

Thalassionema

Rod-like cells united into zig-zag or star-like colonies. Numerous small chloroplasts (**6**.3).

Licmophora

Triangular cells attached by the pointed ends to algae, hydroids or even to planktonic crustacea etc. (**6**.6).

Navicula

The cells creep about in a characteristic jerky fashion. A few are perhaps truly planktonic and some, e.g. *N. membranacea*, are common in most coastal waters. The cells of this species join to form chains.

Pleurosigma

Although true bottom-dwellers, able to creep about and never (?) forming chains, some species are found in inshore plankton samples, particularly after rough weather which stirs up the bottom (**6**.4).

Bacillaria

Bacillaria paradoxa (= *paxillifer*) has very long (up to 250μ), narrow cells which stick together to form colonies of various sizes and which look like a pile of matches. Once seen alive it can never be forgotten, for the cells slide over one another quite rapidly, being at one time almost end to end and at another neatly arranged side by side (**7**.1).

Nitzschia

N. closterium is the commonest species. Its spindle-shaped cells have a long spine at each end. They can move but do not join up as do those of some other species (**7**.2).

IV
Zooplankton

Introduction

AS MENTIONED previously, practically every major group of animals has its representatives in the zooplankton either as adults or as larvae. The division between zooplankton and phytoplankton is taxonomically shadowy, for some flagellates are autotrophic, others are holozoic, whilst even one and the same species may on occasion feed either as an animal or as a plant. Yet in the main a division into phyto- and zooplankton coincides with taxonomic differences of a major kind, so that it is convenient to neglect the border-line nature of some organisms and to deal with the zooplankton group by group in the order in which they appear in most standard textbooks.

Phylum PROTOZOA

Sub-phylum SARCOMASTIGOPHORA

Super-Class CHROMONADEA

Class CHRYSOMONADEA

Order DINOFLAGELLIDA

The dinoflagellates form an order whose typical members are characterized by having two flagella, one trailing along in a groove along the main axis of the cell, the other along a transverse groove. Usually the cell is encased in an ornamented cellulose theca made of two or many plates, but in the simpler kinds this is lacking and the two flagella arise from one end of the cell. There is often an eye-spot or even a simple eye with a lens. The nucleus is large and its 'chromatin' appears to be arranged in beaded filaments. Usually there is a system of vacuoles - probably for the intake of food by means of pseudopodia put out from the ventral (longitudinal) groove, for, although most dinoflagellates can photosynthesize, most are mixotrophic, that is they can also ingest solid food. Some, indeed, are obligate parasites. Many dinoflagellates are armed with complex trichocysts.

Reproduction is usually by binary fission, but in some the daughter cells remain united for some time to form chains. In others, like *Gymnodinium* (= *Pyrocystis*) *lunula*, multiple divisions within a spherical cyst yield up to sixteen sickle-shaped secondary cysts whose contents divide to form up to eight free-swimming individuals again (**9**.1).

Dinoflagellates are classified into two sub-orders, Adinida and Dinifera. Thecate and athecate species occur in both orders but the vast majority belong to the Dinifera, the Adinida being the simpler forms with the flagella springing from one end. Only thecate Adinida occur in the sea.

Some of the commoner dinoflagellates are illustrated in **9-11** and a few notes on some of them follow.

Sub-order ADINIDA

Prorocentrum

Prorocentrum micans (**8**.4) is a widely distributed species, perhaps particularly common in neritic and estuarine waters. *P. scutellum* (**8**.6) is somewhat similar.

Exuviella

Several species are widely distributed both in neritic and offshore waters (**10**.4).

Sub-order DINIFERA

This sub-order includes all the dinoflagellates considered as typical, i.e. from the present point of view all those most likely to be met with in the plankton. Even here there are wide differences in organization.

Gymnodinium

G. lunula (= *Pyrocystis lunula*) is a cosmopolitan neritic species part of whose life-history is shown in **9**.1.

G. splendens may be met with in coastal and estuarine waters (**9**.3).

G. rhomboides (**9**.4), also mainly neritic and estuarine, is especially likely to be found in spring and early summer. Several other species also occur in British plankton.

Amphidinium

Several species can be expected to occur in British waters. One of the commonest is *A. crassum*, which is inshore and widely distributed (**9**.5).

Gyrodinium

G. glaucum (**10**.3) is one of the commonest representatives of this genus but others can be expected to occur more or less frequently.

Polykrikos

Dinoflagellates with a thin pellicle but united to form permanent colonies, each member of which is rather like a *Gymnodinium* so that a composite dinoflagellate with several nuclei (less than the number of individuals as indicated by the number of flagella) is formed.

P. schwarzi (**9**.2) is widely distributed and common in the summer in British waters.

Phalacroma

Species of this genus may be encountered. Many are reported as sand-dwelling forms but some are pelagic (**10**.1).

Dinophysis

Many species are widely distributed around our coasts. One of the commoner ones is *D. acuta* (**10**.2).

Noctiluca

The sole species is *N. scintillans* (=*miliaris*) (**8**.10), a veritable giant among dinoflagellates, and one which is sometimes classified as a separate order of flagellates. It is immediately recognizable as a gelatinous sphere 1mm. or more in diameter. From a ventral depression springs a rudimentary trailing flagellum, whilst from the mouth arises a stout tentacle. There are no chloroplasts, for nutrition is entirely holozoic. *Noctiluca* often occurs in dense swarms, particularly near the coasts, and is one of the main organisms responsible for bioluminescence in temperate waters. It can occur in great numbers well up estuaries.

Goniaulax

Species of *Goniaulax* (**10**.5) are widely distributed and some occur in British waters. Some are luminescent. Together with some species of *Gymnodium*, *Goniaulax* is one of the best-known examples of organisms causing 'toxic blooms' which may be so dense and virulent as to poison all organisms in the food chain which depend on the plankton. *G. lantanella*, for example, is such a one and colours the sea red off the Californian coast.

Diplopsalis

D. lenticula (**10**.11) is not uncommon in the N. Sea and Channel. It is somewhat similar in structure to the true perditions.

Peridinium

The true perditions belong here. They are one of the commonest of all genera of dinoflagellates and are very important in the productivity of the sea. All are more or less top-shaped cells which sometimes bear projecting horns, but these are never as long as in species of *Ceratium*.

P. excentricum (**18**.8) and *P. granii* (**10**.9), although widely distributed, are more typical of colder waters. The most widely distributed common species is probably *P. depressum* (**10**.6). *P. divergens* (**10**.7) is also widely found. *P. globulum* (**10**.10), although found in the Channel, is more typical of warmer seas.

Ceratium

This genus outrivals *Peridinium* in the number of important species it contains. All species are rather flattened cells with an ornamented theca drawn out into one apical and two lower horns.

C. candelabrum (**10**.12), *C. extensum* (**11**.3) and *C. setaceum* (**10**.14), as well as a few others which occur in British plankton, are mainly warm-water species. The rest illustrated in Plates **10** and **11** are temperate or cold-water species.

C. fusus (**11**.5) is one of the commonest of our bioluminescent organisms. *C. longipes* (**11**.6), *C. macroceros* (**11**.2) and *C. bucephalum* are used as indicators of northern water contributing to the S.W. Dagger Bank swirl.

Super-class ARCADIAN

Class RHIZOPODEA

Order FORAMINIFERIDA

Foraminifera are rhizopods with branching and anastomosing pseudopodia which protrude through and form a network over the shell or test. In the vast majority the test is calcareous and perforated by minute pores. It may be a single shell, but most often is built of many chambers. Most of the many-chambered foraminifera occur in two distinct forms: megascleric, with a large initial chamber, and microscleric, with a small initial chamber. The contents of the microscleric forms undergo repeated divisions to form uninucleate young which are released and secrete the initial chamber of a megascleric form, later chambers being added successively. The megascleric forms undergo repeated divisions to form swarms of biflagellate gametes which fuse in pairs to form a zygote. Each zygote then secretes the initial chamber of a microscleric form. Many, but by no means all, foraminifera are marine and planktonic. They are particularly abundant in warmer seas on whose floor tests of dead ones often accumulate to form deep layers of calcareous ooze.

Globigerina bulloides (**15**.2)

This is a cold-water spiny species occurring in sub-polar and temperate seas. Characteristically there are 4-5 spherical chambers in each whorl and the aperture is umbilical in position. The diameter reaches 800μ.

Globigerina pachyderma (**15**.4)

This species is spherical in shape and reaches only 450μ in diameter. There are five rounded chambers per whorl in the juvenile but four more angular chambers per whorl in the adult. The aperture is a narrow slit and bears an obvious lip and there are no spines present in the adult. A sub-polar and cold temperate form.

Globigerina quinqueloba (**15**.6)

The chambers number 5-6 per whorl and are hemispherical in shape. The test is spiny and approximately 270μ. in diameter. The final chamber often forms a flap-like extension over the central part of the coil. A sub-polar and cold temperate form.

Globigerina inflata (**15**.1)

The test is flat on the spiral side but convex on the side of the aperture and reaches 650μ in diameter. There are five hemispherical chambers per whorl in the juvenile but four in the adult. The test is characteristically smooth with a large aperture. A cold temperate form.

Order RADIOLARIA

These are rhizopods with rather stiff, radiating pseudopodia and with the cytoplasm of the body composed of a central mass (enclosed in a perforated capsule) and a more peripheral extra-capsular layer. A nucleus or, more commonly, many nuclei, lie in the central mass. A skeleton formed of silica, which may be merely radiating spicules, but which may be a complex lattice-work, is usually present. Radiolarians thrive in colder seas and their skeletons often form siliceous ooze on the sea bed. They are not common in British seas. Examples are shown in **14**.7-9.

Order ACANTHARIA

The acantharians are distinguished from the radiolarians by having spicules of strontium sulphate ($SrSO_4$) which meet at the centre of the protoplasm and which are regularly arranged in the same pattern in all genera. There is no perforated membrane between the central and peripheral cytoplasm.

Acanthochiasma fusiforme (**15**.3)

The length of the spicules is 400-600 μ and approximately 3 μ thick projecting from a protoplasm of 160-250μ. diameter. A widespread species occurring in the North Sea, east coast of North America and Sargasso Sea as well as in the Mediterranean.

Acanthochiasma serrulatum (**15**.5)

The flattened spines are 550-600μ. long and serrated along each side. The protoplasm is 50-60μ in diameter and has an obvious ectoplasmic and endoplasmic region. Found in the western Atlantic and Sargasso Sea.

Subphylum CILIOPHORA

Class CILIATEA

Sub-class SPIROTRICHIA

Order OLIGOTRICHIDA

Strombidium sp. (**14**.1, 4, 6)

As in all oligotrichs, the cilia are either reduced or bristle-like. There is usually a prominent girdle of trichocysts which may spiral several times round the organism. Most of the species are euryhaline North Sea, Baltic and Atlantic forms.

Tontonia gracillima (**14**.3)

This is an Atlantic form approximately 50μ long and with an elongated tail-like projection some 5-6 times the body length.

Strobilidium sp. (**14**.2, 5)

Small oligotrichs occurring mainly in the Atlantic. Most species are between 100-150μ in length.

Order TINTINNIDA

Tintinnids are mainly marine spirotrichs which secrete a vase-like shell or lorica into which the body can be withdrawn. The lorica is formed of hardened protein and its surface is often covered by diatoms, sponge spicules or other small particles. The body of the animal is often bell-shaped and attached to the lorica by a short oblique stalk. Complicated ciliated membranelles surround the mouth and form the feeding apparatus, together with a series of contractile tentacles. Tintinnids are widely distributed in open seas and coastal waters and many forms show a good deal of variation in the size and proportions of the lorica. Reference should be made to the work of Kofoid and Campbell (1929, 1939), Campbell (1942) and Marshall (1969) for the detailed identification of tintinnids. Some of the more common species occurring in Atlantic and European waters are given below.

Leprotintinnis pellucidus (**12**.4)

The lorica is an open ended cylinder of approximately 200 μ length and 40μ width with a slight constriction near the aboral end. The walls of the lorica are soft and particles tend to adhere to it. A widespread species ranging from the Arctic down through the Baltic and European coastal waters to the coasts of N.W. Africa.

Tintinnopsis acuminata (**12**.1)

The lorica is small reaching only some 80μ in length and 20μ in width. It is closed at its aboral end and has few particles attached to it. The wall of the lorica has a characteristic structure of minute alveoli. This species occurs from the north of the British Isles and Norwegian Sea through the Baltic, North Sea and European coastal waters including the Bristol Channel and Irish Sea.

Tintinnopsis campanula (**12**.7)

The lorica has a widely open trumpet-shaped oral end and often has annulations on it. The aboral end is closed and is prolonged into a pedicel. The species reaches some 200μ in length and up to 150μ width and occurs from the Norwegian coast through the North Sea into the English Channel to the coasts of N.W. Africa; it is also to be found in the north Atlantic extending westwards as far as Nova Scotia.

Tintinnopsis lobancoi (**12**.6)

The lorica is of variable form but is usually long and cylindrical with a bluntly pointed aboral end. The size also varies greatly from 100-400μ in length and 30-60μ in width. There may be many subspecies of this animal which extends from the Arctic seas down to the coasts of N.W. Africa and also in the northern Atlantic across to Nova Scotia.

Tintinnopsis strigosa (**12**.10)

This species may be a subspecies of *T. lobancoi* but the lorica is shorter and has a short and rather broad pedicel aborally; the overall length rarely exceeds 80μ. and the width 40μ. It is a species characteristic of European coastal waters from the Baltic to Biscay.

Stenosmella nivalis (**12**.8)

This is a small tintinnid with a lorica of 30-60μ in length and approximately 15-20μ in diameter. The lorica is characteristically flask-shaped with polygonal markings and attached foreign particles but the collar is smooth and without particles. The species is typically a northern Atlantic form which also occurs from Arctic waters down to the coasts of N.W. Africa.

Stenosomella ventricosa (**12**.9)

This is a much larger species than *S. nivalis*. The lorica is extensively covered with particles and reaches 100μ in length and 40μ in width. The collar is less obvious and the lorica swells out sharply at the oral end. The distribution is very similar to that of *S. nivalis*.

Codonellopsis ecaudata (**13**.11)

The collar is rather longer than the bowl and is marked with 11-13 spiral grooves. The overall length of the lorica is approximately 100μ and the width 35-45μ. Characteristically occurs in the North Sea and English Channel.

Codonellopsis pusilla (**13**.9)

A very small tintinnid with a lorica of only 50-60μ in length and 15-20μ in width. As in *C. ecaudata*, the collar has spiral markings on it but is shorter than the main bowl of the lorica which bears hexagonal markings. The species is found in North Atlantic and Arctic waters as well as in European coastal waters as far south as Biscay.

Dictyocysta dilatata (**13**.5)

The collar of this species is approximately a quarter of the overall length of the lorica and is characterised by a single row of eight openings or fenestrae. The overall length of the lorica reaches 60-70μ and the width 40-50μ. The bowl of the lorica is also penetrated by fenestrae which are arranged in seven rows. A North Atlantic species occurring also in European coastal waters.

Dictyocysta elegans (**13**.7)

In this species the collar is longer than in *D. dilatata* and is penetrated by two rows of fenestrae, the eight oral ones of which are larger than the ten basal ones. The lorica is also penetrated by three whorls of fenestrae and reaches 60-70μ in length and 40-50μ in width. A widely distributed form occurring in the North Atlantic as well as in Arctic and European waters as far south as the coasts of N.W. Africa.

Codonella amphorella (**13**.8)

This species is distinguished from *Tintinnopsis* by the fact that there is an obvious division of the lorica into collar and bowl. But the collar is not marked by spirals or rings and never longer than the bowl of the lorica. *C. amphorella* reaches 80-100μ in length and 40-50μ, width. The pedicel is pointed and sealed from the main part of the bowl of the lorica. A North Atlantic species occurring also in the North Sea.

Coxiella ampla (**13**.10)

The cup-shaped lorica has no collar and no foreign particles attached to its surface; there is a spiral groove running from the closed aboral end to the oral end. The overall length is up to 200μ and the width 60-90μ. A widely-distributed tintinnid which is found in Arctic waters down through the North Sea to the coasts of N.W. Africa as well as in the central North Atlantic.

Helicostomella subulata (**13**.3)

This genus may contain many subspecies. The lorica is characteristically elongated and ends aborally in a slender pointed pedicel. The oral end has a variable number of spiral markings. The overall length reaches 200-500μ and the width 20-25μ. This species is widespread and is found in Arctic waters as well as in the North Sea, English Channel and in the Atlantic from the coasts of N.W. Africa to Newfoundland and Nova Scotia.

Flavella ehrenbergii (**13**.2)

A very large tintinnid with a lorica which reaches 400μ in length and 125μ in diameter. The lorica is cylindrical with some oral spiral markings and with a short blunt pedicel which bears flattened buttress-like structures joining it to the main cup of the lorica. Distributed in the North Sea from Norwegian waters through the English Channel to the coasts of N.W. Africa.

Ptychocylis arctica (**13**.5)

The lorica is characterized by two bulges, one of which is near the serrated oral end and the other a little further back. The overall length reaches 150μ and the width 100μ. A northern species extending from Arctic waters to the North Sea as well as from the northern to central North Atlantic regions.

Acanthostomella norvegica (**13**.12)

A very small rather variable lorica which is characterized by two collars around the mouth with a gutter running between them. The inner collar is short and upright but the outer collar protrudes outwards and bears a number (usually 20-35) incurved teeth. The overall length is up to 50μ and the width 25μ. A northern form extending from Arctic waters southwards into the English Channel and from the northern to central North Atlantic regions.

Petalotricha ampulla (**12**.2)

The lorica of this genus bears obvious similarities with *Acanthostomella* but is much larger and reaches 160μ in length and 135μ in width. The outer lip flares outwards and its rim is often serrated. There are window-like apertures or fenestrae at the top of the inner lip as well as a double row on the main bowl of the lorica. A northern North Atlantic species occurring also in the English Channel.

Rhabdonella amor (**13**.4)

The lorica of this species is distinguished by the presence of a series of spiral ribs which branch and anastomose near the middle of the lorica. There is also a large number of small fenestrae perforating the wall of the lorica. The overall length reaches approximately 100μ and the width is generally approximately 50 μ. Characteristically a northern Atlantic form extending as far south as European coastal waters and the Azores.

Epiplocyloides reticulata (**13**.1)

The lorica is 60-70μ in length and 45-50μ in width. It is reticulated over the aboral surface whilst the oral half is covered with fine grooves extending towards the collar. This species is found in the English Channel and is possibly a variety of a species found in the central North Atlantic region.

Parafavella denticulata

The lorica is very large, reaching 325μ in length and 60μ in width; it is toothed round the oral margin and ends in an obvious pedicel. The surface is covered with reticulations. A very widely distributed northern species occurring in the northern North Atlantic and Arctic seas down through the North Sea to the coasts of N.W. Africa.

Salpingella acuminata (**12**.5)

The lorica is long and tubular with an open aboral end and reaches 370μ in length. The aboral end bears 6-9 curved spiral fins and the oral end flares widely. A very widely distributed Atlantic species occurring also in Arctic and North Sea waters as well as in the English Channel.

Steenstrupia steenstrupii (**12**.3)

The lorica is very similar in form to *Salpingella acuminata* but reaches only 160μ in length and 50μ in width. The aboral fins are vertical, however, and only six in number. A north and western North Atlantic form extending also from Norwegian waters southwards through the North Sea.

Phylum COELENTERATA

Class HYDROZOA

Most of the familiar hydrozoans are polymorphic coelenterates which form colonies of zooids. These may be either of the medusoid or polypoid type of individual or of both. Some hydrozoans occur solely in the hydroid or medusoid phase. By far the commonest Hydrozoa found in the British plankton are the medusae (pelagic reproductive zooids) of the bottom-dwelling colonial zoophytes belonging to the order Hydroida; but medusae which are holoplanktonic, that is, are not related to a fixed colonial stage (Trachylina), as well as colonies of mixed hydroid and medusoid types (Siphonophora), will also be encountered, particularly in warmer waters off our western coasts and in the northern N. Sea.

Order HYDROIDA

The vast majority of British hydrozoans belong here. The fixed, bottom-dwelling or hydroid stages are usually polymorphic and colonial. Most release pelagic reproductive zooids, called medusae, which bear the male or female gonads. In many, however, the medusoid stages are retained on the parent colony as gonophores. These need not concern us here. It is at once obvious that a hydroid may have been first encountered either as a fixed colonial stage or as a free-swimming medusa and it is still true today that many medusae have not been related with certainty to the fixed stage and *vice versa*. So whilst, ideally, only one set of generic and specific names should be given to the polymorphic stages of a single species, in practice a dual nomenclature may for a time prevail. For example, *Phialidium hemisphericum* was described first in 1760 and called *Medusa hemispherica*, but it was transferred to various genera by subsequent workers. It was later shown to be the medusa of the hydroid *Clytia johnstoni*. According to the International Rules of Nomenclature the name *Phialidium hemisphericum* should take priority. In practice planktonologists tend to retain the name first given to the medusoid stages of any of the Hydroida. Moreover, in this context it is more convenient to speak of Anthomedusae than Gymnoblastea, or of Leptomedusae than of Calyptoblastea, although the two sets of names mean the same thing taxonomically, the second named in each of the two sets of terms referring to the fixed colonial phase and the first to the medusoid phase.

The main divisions of the British Hydroida (orders or sub-orders, according to different authorities) are five in number: Anthomedusae (Gymnoblastea), Leptomedusae (Calyptoblastea), Limnomedusae, Trachymedusae and Narcomedusae. Some notes on each of these sub-orders now follow. Russell (1953) and Kramp (1961) should be consulted for authoritative accounts.

Sub-order ANTHOMEDUSAE (Gymnoblastea)

These are the medusae of the gymnoblast hydroids, distinguished by having polyps not retractile into cups or hydrothecae. The medusae are bell-shaped and bear gonads on the stomach or manubrium. As a rule the tentacles around the margin of the bell are few. The sense organs are eye-spots (ocelli). Statocysts are absent.

Sarsia eximia (**16**.1)

This is the medusa of the hydroid *Syncoryne eximia* and is mainly confined to coastal waters, although widely distributed from April to September around our coasts from the North Sea through the English Channel to S.W. Ireland. It is a fairly large and active medusa even when first liberated from the hydroid stage, finally reaching a height of about 4mm. The stomach does not extend beyond the margin of the umbrella and there are no buds formed asexually.

Sarsia prolifera (**16**.1)

This medusa has not been related to any hydroid but is presumably the medusa of some species of *Syncoryne*. It differs from *S. eximia* most obviously when it has reached the stage at which secondary medusae are produced asexually by budding at the base of the tentacles. When fully grown it is only about 2mm tall. It occurs in the plankton from June to October in the North Sea, English Channel and Atlantic off S.W. Ireland but is primarily a south western species.

Sarsia tubulosa (**16**.6)

Probably the medusa of *Syncoryne decipiens*; differs from the two previous species in having a long tubular stomach and in growing much larger, sometimes attaining a height of about 10mm. in British waters. It occurs in its greatest abundance in spring (April), and is only rarely caught after the end of June. It is widely distributed in coastal and estuarine waters from the Norwegian Sea, Baltic and North Sea to the coast of S.W. Ireland.

Sarsia gemmifera (**19**.5)

This has not been related to a hydroid. It is distinguished by its ability to produce buds from the wall of the manubrium which can extend beyond the margin of the umbrella and is dilated at its end. The bell can reach a height of 5mm. The ocelli are black and the tentacles short. It is found all round our coasts from the Baltic to S.W Ireland in summer and early autumn but is not common.

Dipurena halterata (**19**.6)

This is a large medusa which reaches 8mm in height and has four marginal tentacles each bearing a brownish-red terminal cluster of nematocysts as well as 3-6 rings of nematocysts near the tip of each tentacle. The manubrium is very long and bears the gonads in two or three regions giving a segmented appearance. The medusa occurs in the North Sea, English Channel and waters off S.W. Ireland and can be distinguished from *Sarsia tubulosa* by the swollen mouth region of the manubrium and the terminal swelling of nematocysts on each tentacle.

Dipurena ophiogaster (**19**.7)

This differs from *D. halterata* in the absence of rings or terminal clusters of nematocysts on the tentacles. Instead the nematocysts are scattered along the length of the tentacles. The manubrium is very long and bears the gonads in up to six principal segments. The overall height is rather less than in *D. halterata* and reaches only 5mm. A south western species occurring in the western approaches to the English Channel and in the waters off S.W. Ireland.

Hybocodon prolifer (**16**.8)

The medusa of the hydroid of the same name. It has from one to three marginal tentacles arising close together on the bell margin. There are five meridional tracks of nematocysts on the exumbrellar surface. The stomach is short and broad and actinulae larvae are grown from the manubrium. The overall height of the bell reaches 4mm. It is widely distributed and occurs all around our coasts, particularly in the north, in spring and early summer.

Euphsya aurata (**19**.1)

This medusa resembles *Steenstrupia* in that only one tentacle with rings of nematocysts may be present although in some specimens there may be up to four tentacles. It is distinguished by the fact that the apex of the bell is rounded, not pointed as in *Steenstrupia*. Stomach, marginal bulbs and tentacle are all often yellow with a red tip to the oral end of the stomach; the bell reaches 6mm in height. This species generally occurs in the North Sea, English Channel and off S.W. Ireland from April until November.

Ectopleura dumortieri (**19**.2)

This is a small rather spherical medusa reaching 2mm in height and with eight meridional nematocyst tracts on the exumbrellar surface. There are four brownish yellow tentacles which bear clusters of nematocysts at the tips. The stomach is spherical with tiny red markings round its centre and tapers towards the mouth. Not a common medusa but recorded from the southern North Sea, English Channel, S.W. Ireland and North Atlantic.

Steenstrupia (=*Corymorpha*) *nutans* (**17**.6)

The medusa of *Corymorpha nutans* has, as its most obvious character, a single tentacle on the bell margin and a large yellow or brown stomach, and a pointed apex to the bell. It is a widely distributed inshore species reaching some 6mm in height.

Eucodonium brownei (**16**.2)

The medusa, presumably of a *Tubularia*-like hydroid, is included here as exemplifying the difficulties of relating medusa to hydroid phase. It is rare but has been recorded from Plymouth and from the N. Sea. Asexual medusa buds are formed on the stomach. It is a small medusa, rarely exceeding 1mm in diameter.

Margelopsis haeckeli (**19**.4)

The medusa reaches 2mm in height and has many nematocysts on the exumbrellar surface. The gonad surrounds the upper part of the stomach and actinulae larvae develop from its surface. There are four marginal bulbs which may have black pigmentation and have 3-4 tentacles arising from the surface of each. This species may occur in large quantities during the summer in the southern North Sea.

Margelopsis hartlaubi (**19**.3)

This is a rather larger medusa than *M. haeckeli* reaching 3mm in height and with only two tentacles arising from each of the four marginal bulbs. Each tentacle has 4-6 rings of nematocysts. A northern North Sea species.

Cladonema radiatum (**16**.4)

This is the medusa of the hydroid of the same name. It has branched marginal tentacles which bear clusters of nematocysts. The base of each tentacle is thickened and bears characteristic adhesive organs which are usually three in number in the adult. The stomach has five pouched outgrowths. The five radial canals may branch to form secondary canals making a total of eight in all, since some remain unbranched. The ocelli are very obvious. It is a most active medusa and is widely distributed, occurring in Skagerak and southern North Sea areas but principally to the south west of the British Isles.

Zanclea costata (**19**.8)

This is the medusa of the hydroid of the same name. It has four exumbrellar patches of nematocysts. There are four marginal tentacles, which are very extensible, and bear numerous stalked oval capsules each containing 2-5 nematocysts. It is a widely distributed but not very common species occurring all round the British Isles as well as in the Atlantic and Pacific. In British waters it is found mainly from May to September.

Turritopsis nutricula (**17**.2)

Probably the medusa of the hydroid *Turritopsis* (=*Dendroclava*) *dohrnii* which occurs on the east coast of North America and in the Mediterranean. The hydroid has not yet been identified in British waters but the medusa is common in summer and winter in the English Channel and North Sea; it is characterised by a single row of 80-90 marginal tentacles around the margin of the bell which reaches 4-5mm in height. The manubrium does not extend beyond the bell and the mouth has small rounded groups of nematocysts along the margins of the four lips. There are also characteristic vacuolated cells above the stomach which, like the gonads, is bright red in colour.

Rathkea octopunctata (**19**.9)

The medusa of a minute hydroid of the same generic name which has been identified so far only from specimens reared in the laboratory. It reproduces by asexual budding from the wall of the manubrium which has a mouth bounded by four lips each divided into two short branches bearing clusters of nematocysts.

There are eight groups of marginal tentacles with no ocelli at their bases. There are four radial canals and the gonads completely surround the stomach. The number of marginal tentacles increases with age until there are five in the per-radial positions and three in each inter-radial position. The medusa reaches a height of 4mm. and occurs all around our coasts, particularly in winter and early spring. It is also found in Arctic waters, the Mediterranean and western North Atlantic area. The young can be confused with *Lizzia blondina* adults, from which they are best distinguished by the position of the origin of the oral tentacles. In *Rathkea* these arise strictly at the mouth margin and are branched. In *Lizzia* the bases are above the mouth and the oral tentacles are unbranched.

Lizzia blondina (**19**.10)

This is the medusa of an unknown hydroid. It reproduces by asexual budding from the manubrium and in this and other ways closely resembles *Rathkea octopunctata* (see above). There are four groups of three perradial tentacles alternating with single interradial tentacles. It reaches 2mm in height and is common in the plankton all around our coasts except in the eastern part of the Channel and southern N. Sea. Seasonally it is most abundant from May to October.

Bougainvillia ramosa (**16**.5)

The medusa of the hydroid bearing the same name. Despite the wide occurrence of the hydroid, the medusa has seldom been recorded in British waters. It seems to be mainly a summer and southern species. The most distinguishing feature is the presence of four branched tentacles around the mouth; *B. britannica* and *B. principis* are somewhat similar one to another and to *B. ramosa.* Only attention to those details given in Russell (1953) will distinguish the species, but the first two mentioned have a more northerly distribution than *B. ramosa.* Apart from these, other species of *Bougainvillia* occur in western waters.

Leuckartiara octona (**19**.11)

This is the medusa of the hydroid of the same name, formerly known as *Perigonimus repens.* It is a very common medusa all around our coasts and occurs also in Arctic waters, as well as in the Mediterranean and in the western North Atlantic. The bell may reach a height of 15mm and can be distinguished by the very folded margins to the four lips. The gonads are horseshoe-shaped with folds directed outwards and there are up to 23 marginal tentacles. It can be regarded mainly as a summer and early-autumn species. Other, but rarer, species are recorded in British waters.

Neoturris pileata (**18**.2)

This is rather a large medusa reaching 35mm in height and characterised by as many as 90 marginal tentacles. The gonads have complex inwardly directed folds.

A deep-water species occurring in the western English Channel, S.W. Ireland and to the west and north of the British Isles in general.

Sub-order LEPTOMEDUSAE (Calyptoblastea)

These are the medusae of the calyptoblast hydroids which have polyps fully retractile into hydrothecae. Although sometimes bell-shaped, the medusae are usually more flattened or saucer-like than Anthomedusae and, moreover, have their gonads situated on the radial canal (not on the manubrium). There are usually numerous marginal tentacles which increase in number throughout life. The sense organs are marginal vesicles or statocysts - not ocelli (which can occur but only in a few examples). Marginal clubs or cordyli replace vesicles or ocelli in the Laodiceidae, whilst in the Melicertidae and Dipleurosomidae sense organs are lacking from the margin of the bell.

Laodicea undulata (**17**.3)

The medusa of a species of *Cuspidella.* Its main distinguishing features are the broad velum; the four-sided stomach; the mouth with four lips with folded margins; the 300 or more of hollow marginal tentacles some of which bear small ocelli; and cordyli between each pair of tentacles. It may attain a diameter of 15mm or so. Widely distributed around our coasts, it occurs most commonly in west, south-west and north areas, particularly in late summer and autumn.

Staurophora mertensi (**18**.8)

This is a very large leptomedusa which may reach 300mm in width. It is a cold-water species occurring to the north and west of the British Isles and is occasionally found in the northern North Sea. There are four radial canals with many short diverticula which bear the gonads and up to several thousand short hollow marginal tentacles. The medusa differs from *Aequorea* which has many radial canals rather than only four.

Melicertum octocostatum (**17**.1)

The medusa of the hydroid of the same name. Three to seven tracts of nematocysts on the exumbrellar surface are important characters. There are no marginal sense organs but 64-72 marginal tentacles in the adult which alternate with much shorter marginal tentacles. The bell may reach a diameter of 14mm. Although it has a wide distribution, it is mainly a northern species.

Cosmetira pilosella (**17**.4)

This is the medusa of a species of *Cuspidella,* probably *C. grandis.* There are up to 100 marginal tentacles with swollen bases. There are eight marginal vesicles but no ocelli. The gonads on the radial canals are long. Although occurring from the Channel to the N. Sea, it is mainly a S.W. species, most abundant in the summer. Up to 48mm. in diameter.

Mitrocomella brownei (**18**.3)

The bell of this species reaches 7mm in diameter and there are four radial canals which bear oval gonads near their ends. There are 16 marginal tentacles and 8 marginal vesicles whereas in *Mitrocomella polydiademata* there are double the number of tentacles and vesicles; the gonads are also much longer than in *M. brownei.* The latter is a south-western species occurring during the summer and early autumn in the western English Channel, S.W. Ireland and to the west of the British Isles whereas *M. polydiademata* occurs to the north of the British Isles during the spring and early summer.

Tiaropsis multicirrata (**18**.6)

This is a northern species occurring to the north and west of the British Isles as well as in Arctic waters and in the northern North Atlantic. The bell reaches 20mm diameter and bears 4 radial canals with gonads developed along their central region. There are up to 300 marginal tentacles and eight marginal vesicles. Each vesicle has a pigmented ocellus which distinguishes this medusa from other leptomedusae.

Phialidium hemisphericum (**18**.7)

The medusa of *Clytia johnstoni.* It is one of the commonest British medusae and has been reported from all round the British Isles, most abundantly during the spring and autumn. It may attain a diameter of 20mm but is normally 5-10mm. There are thirty-two or so hollow marginal tentacles and up to three marginal vesicles are to be found between each pair of tentacles. There are four radial canals and the gonads are situated near the end of them but not quite reaching the peripheral ring canal.

Obelia sp (**17**.5)

These are the medusae of hydroids of the genus *Obelia.* They are well known from descriptions in various textbooks and are characterized by having a very flat bell; rounded gonads on each of the four radial canals; numerous solid marginal tentacles; short manubrium; absence of ocelli; and eight adradial marginal vesicles. Although numerous species of hydroids are distinguished, their medusae are difficult to identify down to species. The medusae are common from spring to autumn in the plankton but may be found at other times of the year.

Phialella quadrata (**17**.7)

The medusa of the hydroid of the same name, formerly known as *Campanulina repens.* The bell is nearly hemispherical and has around its margins up to thirty-two tentacles. There are no ocelli, but there are eight marginal vesicles in the adradial position. The bell may reach a diameter of 13mm. The medusa is common around the British Isles, being particularly characteristic of the western part from Scotland to the W. Channel. It is less common off our eastern coasts.

Eucheilota maculata (**18**.4)

A southern and central North Sea medusa which reaches 13mm diameter and 10mm height. There is a characteristic black spot on each interradial wall of the stomach and there are spiral marginal cirri on each side of the bases of the marginal tentacles which are 20 or more in number.

Aequorea vitrina (**20**.3)

Probably the medusa of *Campanulina acuminata.* Species of *Aequorea* are immediately recognizable, even when partially grown, by their immense size. When of full size they outrival many proper jellyfish (Scyphozoa), although their hydroids are insignificant. *A. vitrina* can reach a diameter of 170mm. It has a flattened bell bearing up to 600 marginal tentacles and at all stages of growth these are more than three times the number of the radial canals. The gonads are in thin strips on each side of the radial canals. The mesogloea is thick and the velum narrow. The large stomach has a diameter half that of the umbrella - usually somewhat less. The margins of the mouth are drawn out into tentacle-like projections with crenellated margins. What appear to be canals, open on the ventral surface, radiate from approximately every alternate radial canal to the mouth margins.

It seems from the literature that *A. vitrina* is not a common or widely distributed species, yet each year from mid-June to the end of July these medusae appear in large numbers at Whitstable and it is difficult to believe that they are not common in other localities.

A. aequorea is probably the medusa of *Campanulina paracuminata.* Although similar in structure and size to *A. vitrina*, it can be distinguished by the fact that it has fewer radial canals and that its marginal tentacles are only about the same or, more rarely, twice these in number, perhaps never exceeding 120. *A. aequorea* has a more restricted distribution than *A. vitrina*, being apparently a more southerly and S.W. species. *A. macrodactyla* has a similar distribution. It has from 120 to 180 marginal tentacles and these are much fewer in number than the radial canals.

Eutima gracilis (**20**.2)

The hydroid is not known with certainty, but probably resembles *Octorchis gegen-*

bauri. The nearly hemispherical adult has a thick mesogloea. Only two, or sometimes four, marginal tentacles. Common in the S.W. but also recorded in northern N. Sea, and Mediterranean.

Tima bairdi (**20**.1)

Probably the medusa of a small campanularian hydroid, but this has not been determined with certainty. There are sixteen marginal tentacles when the medusa is fully grown. The mesogloea is very thick and the four-sided manubrium is small. The gonads are folded and linear and there are 200-250 marginal swellings between the marginal tentacles. *Tima bairdi* has been recorded only from the east coasts of the British Isles, but is rare south of the Tyne or Tees, so that it can be regarded as a good indicator of northern N. Sea water in which it is sporadically abundant.

Sub-order LIMNOMEDUSAE

These are Hydromedusae which either have gonads on the wall of the manubrium or along the radial canals and in which marginal vesicles may or may not be present. There are no bulbous swellings at the base of the marginal tentacles. In a few the gonads may occur both on the manubrium and on the radial canals. From this it is apparent that the characters of the Limnomedusae overlap those of the Anthomedusae and the Leptomedusae. In some the statocysts take the form of enclosed sensory clubs and in this respect the Limnomedusae resemble the Trachymedusae. There are only a few marine Limnomedusae, most being freshwater.

Only two species are likely to be found in British waters, *Proboscidactyla stellata* and *Gossea corynetes.* Both are mainly S.W. species.

Proboscidactyla stellata (**18**.5)

This medusa reaches 9mm in width and has been recorded in spring and summer from many localities to the north and west of the British Isles as well as in the Thames estuary and Norwegian waters. There are six primary radial canals each of which bifurcate twice to give 24 canals at the periphery of the bell. The stomach is six-lobed each of which extends for some distance out over the sub-umbrellar surface. The gonads extend out over each lobe of the stomach. There is no ring canal and there are 24 marginal tentacles.

Gossea corynetes (**18**.1)

A southern species occurring in the southern North Sea, English Channel, off S. W. Ireland and in the Mediterranean. The medusa reaches 15mm in width and 107mm height and the gonads form deep folds hanging below the four radial canals. There are eight groups of three solid marginal tentacles with one or two small tentacles between each group of large ones. The large tentacles characteristically bear rings of nematocysts and there are usually 24 marginal vesicles on the exumbrellar surface of the ring canal.

Sub-order TRACHYMEDUSAE

The Hydromedusae belonging to this sub-order are characterized by a practically hemispherical shape with a thick nematocyst ring, and by gonads usually confined to the margins of the radial canals. The marginal tentacles are solid in some species but hollow in others. There is no hydroid stage. Trachymedusae are separated from the Narcomedusae in that the margins of the umbrella are not lobed. There is a large muscular velum. The marginal tentacles are easily detached. Trachymedusae are scarce in British waters and all are oceanic, rarely being taken in inshore waters.

Liriope tetraphylla (**16**.9)

The manubrium of this species is very long and bears four lips with clusters of nematocysts along the edges. There are four radial canals with broad flattened gonads below them. The perradial marginal tentacles are long with nematocyst rings around them but the other tentacles are shorter with nematocysts clusters between them. A warm water medusa which may enter the English Channel with water from the south west.

Aglantha digitale (**16**.7)

In the form of the variety var. *rosea* is a trachymedusan characteristic of western waters but may enter the N. Sea from the north and also the S. Bight by way of the Channel.

For details of the structure and distribution of other Trachymedusae see Russell (1953), Kramp (1961).

Sub-order NARCOMEDUSAE

These are medusae with firm, glassy bells with thin sides. Their margins present a scalloped appearance because of thickened ridges of the ectoderm from the inner surface of the tentacle extending on to the exumbrellar surface and joining a thick circumferential tract of nematocysts. The radial canals, so characteristic of most hydromedusans, are absent, as is also the manubrium - the pouched stomach being confined to the interior of the bell, whilst the mouth opens directly into it. Gonads lie on the stomach wall. The solid tentacles arise a little way above the margin of the bell and the sense organs are solid club-like structures. Narcomedusae are oceanic organisms not likely to be met with except in offshore hauls round the western coasts of the British Isles and will not be considered further. They have a direct type of development (planula-actinula-medusa), as in the trachylines, some having a parasitic stage on other Hydromedusae. None of these is known to be British.

Order SIPHONOPHORA

This order contains holoplanktonic hydrozoans, all of which consist of polymorphic colonies built up of several kinds of hydroid and medusoid-like zooids, but none of these ever appear as free medusae, as commonly understood. The hydroid zooids themselves are usually of three sorts: gastrozooids (feeding polyps) with a single hollow tentacle springing from near the base; dactylozooids, also with a single tentacle bearing nematocysts, but lacking a mouth; and gonozooids which may have a mouth but lack a tentacle, and bear clusters of gonophores on branched stalks. The medusoid zooids are usually very different from typical medusae and take the form of swimming bells (also called nectocalyces), bracts (hydrophyllia), or gonophores or floats (pneumatophores). The colonies, formed by asexual budding from a sexually produced planula-like larva, may be compact (as in *Velella* or *Physalia*), or much-branched (as in *Physophora*, *Agalma* etc.). Colonies are rarely recovered intact in plankton hauls.

Siphonophores are, with a few exceptions, uncommon in British plankton and are characteristic mainly of western waters, although some get transported round Scotland into the northern N. Sea. The order falls into two sub-orders, Calycophora and Physophorida.

Sub-order CALYCOPHORA

These are siphonophores with one to many swimming bells at the upper end of the colony. There is no float. The commonest British representatives are the Monophyidae, like *Muggiaea*, which have a single swimming bell (**21**.2). The Diphyidae have two swimming bells and perhaps the monophyids are very closely related to them. Diphyids in British waters include *Lensia* (**21**.7) and *Dimophyes arctica* (**21**.5). *Hippopodius* (**21**.3) is another member of the sub-order possibly to be met with, although it is a more southerly species and is sometimes carried northwards in water of Biscay origin together with species such as *Eudoxides spiralis* (**21**.6) and *Chelophyes appendiculata* (**21**.8).

Sub-order PHYSOPHORIDA

The colony has an apical float, but apart from this feature the colonies are so variable as to warrant considerable taxonomic sub-division. By some, the chondrophorids, like *Velella* (**21**.1), are put into a separate sub-order. *Physalia* (**21**.4), the well-known Portuguese Man-o'-War, is, like *Velella*, an occasional visitor from southern waters to our coasts. *Nanomia cara* and *Agalma elegans* are also indicators of S.W. water (**22**.3 and 4).

Class SCYPHOZOA

Jellyfish in the usual sense of the word. That is, large medusae lacking a velum;

usually free-swimming; with tentacles from the stomach; with sense organs in the form of tentaculocysts; with no hydroid stage or perhaps with a scyphistoma which buds off ephyrae.

These large jellyfish are rarely taken in the ordinary plankton samples and if they are they are so immediately recognizable from descriptions in books readily available that they will not be further described. But mention maybe made of the four common British representatives, namely *Aurelia aurita* (which is a world-wide species), *Cyanea capillata*, *Chrysaora isosceles* and *Rhizostoma octopus*. All four species are seasonally common around our coasts; it is merely their large size and activity which keep them out of most plankton nets.

Phylum CTENOPHORA

In contrast with true coelenterates, the ctenophores are bilaterally, not radially, symmetrical. Moreover they lack nematocysts. Their most obvious feature is the possession of eight meridional rows of ciliated plates or ctenes. Few are members of the British plankton and those which do occur are immediately recognizable. All British ctenophores are holoplanktonic.

In fact only *Pleurobrachia pileus* and *Beröe cucumis* (**22**.6 and 7) are likely to be locally abundant in western waters in some years (**22**.5) although they are rarer, but of larger size, during the winter. *Bolinopsis infundibulum* may be locally abundant in western waters in some years (XXI. 7).

Phylum CHAETOGNATHA

Chaetognaths, or arrow-worms, are active planktonic predators which have a practically world-wide distribution. The characters by which they can be readily recognized in plankton samples are their fairly large size; their elongated torpedo-like shape; their transparency; their paired lateral, and expanded caudal, fins; and their head, bearing a pair of eyes and a series of curved spines around the mouth. In mature specimens the shape and proportions of the body, as well as the position and shape of the gonads and seminal vesicles, are important specific characters.

Only species of the genus *Sagitta* are likely to be met with in British waters, although *Eukrohnia* and *Spadella* have been taken off the S.W. of Ireland and *Eukrohnia* in the region of the Faeroes and northern waters. Indeed, in most sea areas only two species of *Sagitta* are common, *S. setosa* and *S. elegans*, and even these are not usually taken together, *S. elegans* being typical of mixed western and Channel water and *S. setosa* of coastal and Channel water. On the whole *S. elegans* is a northern species, as the Channel marks its southern limit. See also Russell (1939) and Fraser (1957).

Sagitta setosa (**23**.1)

This is an inshore neritic species occurring through much of the North Sea and English Channel and reaches 20mm but is often much smaller than this, especially in the southern North Sea and eastern parts of the English Channel. The characteristic features are that the seminal vesicles are wedge-shaped and situated immediately behind the second lateral fin; the gut is very obvious and without diverticula; the ovaries are rather short and the specimens remain transparent in formalin.

Sagitta elegans (**23**.2)

This chaetognath characterizes mixed oceanic and coastal waters and occurs in the northern North Sea and western Channel approaches. It reaches 25mm in length and becomes somewhat cloudy in formalin; a pair of anterior gut diverticula is present. The body is also somewhat stiffer than *S. setosa* and tends to flip off a needle when attempts are made to lift it out from a sample by this means. The seminal vesicles are conical not wedge-shaped, and are set well back from the second lateral fin towards the tail. There are several sub-species, some of which may exceed 25mm in length.

Sagitta serratodentata (**23**.3)

This is an oceanic species occurring principally to the south west, west and northern regions of the British Isles. The overall length reaches 15-20mm and the seminal vesicles are very large. The body is thin and opaque in formalin and the inner margins of the bristles comprising the jaws are serrated. There are two sub-species, *S. s. tasmanica* which is a northerly form and has a semi-circular anterior edge to each seminal vesicle and *S. s. atlantica* which has two small spines on the tip of each vesicle and is a lusitanean form.

Sagitta maxima (**23**.4)

A very large chaetognath reaching 80mm in length. The body is rather transparent in formalin with a characteristically wide head and oval seminal vesicles. The posterior fin is joined to the anterior one which ends at, or even anterior to, the ganglion. A cold-water species occurring in the deeper waters of the northern North Atlantic.

Sagitta hexaptera (**23**.5)

This is also rather a large species reaching 50-60mm and with a body which remains transparent in formalin. The diagnostic feature is that the anterior pair of lateral fins are small and rounded. The seminal vesicles are rounded and set well back near the tail. An Atlantic species occurring to the north and west of the British Isles.

Krohnitta subtilis (**23**.6)

This is characterized by the presence of only one pair of rather ovoid lateral fins which are displaced posteriorly towards the tail. The body reaches 15mm in length and is slender. A north-eastern Atlantic form.

Sagitta lyra (**23**.7)

This species is transparent in formalin and reaches 40mm; as in *S. maxima* the lateral fins are joined. However, it can be distinguished from *S. maxima* by the normal width of the head and the fact that the anterior fins join the body well posterior to the ganglion. A north-eastern Atlantic form occurring to the west and north of the British Isles and occasionally in the northern North Sea.

Eukrohnia hamata (**23**.8)

The body has one pair of elongated lateral fins joining the body anteriorly to the ganglion. The eyes are unpigmented and the body is opaque in formalin. The overall length reaches some 40mm. A common cold-water species occurring in surface waters in the Arctic but deeper further south.

Sagitta zetesios (**23**.9)

A very wide body, opaque in formalin and reaching 40mm in length. Anterior gut diverticula are present. A deep-water oceanic species occurring around the northern regions of the British Isles.

Sagitta macrocephala (**23**.10)

The head is very large and the gut may be reddish in colour. There are no gut diverticula and no pigment in the eyes. This species occurs to the west and north-west of the British Isles.

Phylum ANNELIDA

Class POLYCHAETA

Polychaete worms are sufficiently familiar to require no general description, but, despite their diversity and abundance in all shallow seas and shores, few are regularly planktonic as adults. In British waters members of only four families of errant polychaetes are likely to be encountered (and these mainly in S.W. sea areas): Phyllodocidae, Alciopidae, Tomopteridae and Typhloscolecidae. All except the tomopterids are considered rare.

On the other hand, many polychaetes are transient members of the plankton even in the adult phase. Even lugworms can and do swim in the surface waters on occasions, whilst many other polychaetes live there for a short time in the breeding season. Particularly is this true of those which become epitokous, and sexual forms or 'epitokes' of various species of *Nereis* and many others (Fage and Legendre, 1927), including scalibregmids (Clark, 1953), may be abundant at the appropriate season. Further, asexual reproductive stages of various syllids are not uncommonly found. For these and other polychaete species Fauvel (1923) is undoubtedly the best general book to consult. Muus (1953) gives excellent keys and figures of holoplanktonic adults met with in the N. Atlantic, whilst Hammond (1967) gives an account of the autolytoid syllids occurring in the plankton.

Lagisca hubrechti (**24**.8)

This scaleworm has up to 46 segments and the dorsal surface is covered anteriorly by 15 pairs of scales or elytra which bear small conical protuberances on the surface. A deep-water species reaching some 20mm and occurring in the northern North Atlantic, to the south-west of Ireland and north-west of the British Isles.

Lopadorhynchus uncinatus (**24**.4)

This is a planktonic Phyllodocid which reaches 10-20mm in length and has 21-32 segments. The body is pale yellow in the living animal and clearly divided into an anterior region consisting of the head plus first two setigerous segments and a more posterior region in which the segments are more slender. A warm-water Atlantic species occurring to the south-west of Ireland.

Greefia celox (**24**.9)

This Alciopid polychaete has approximately 60 segments and reaches 20-60mm in length. There are well-developed red eyes and four pairs of tentacular cirri, as well as a series of pigment spots on the dorsal surface at the base of each parapodium. The pedal lobe has two cirriform appendages. A western and south-western form occurring occasionally in the northern North Sea.

Callizona setosa (**24**.10)

This species also belongs to the Alciopidae and has a thin translucent body of 40-60 segments and reaches some 50-100mm in length. As in *Greefia celox*, there are dorsal pigment spots at the base of each parapodium. There are five antennae and five pairs of tentacular cirri. The pedal lobe has one thin cirriform appendage. The eyes are spherical and brownish in colour. A Mediterranean species occurring also in the lusitanean stream to the south-west of Ireland.

Tomopteris spp. (**24**.5 and 7)

These pelagic polychaetes are quite distinctive with paddle-shaped biramous parapodia which lack setae. There is a pair of nuchal organs on each side of the head. The first pair of setigerous segments are often lost in the adult and the second pair are often very elongated. The species most likely to be encountered are *Tomopteris septendrionalis* which reaches 15-20mm in length and *T. helgolandica* which is nearly double this size. The latter occurs throughout the North Sea as well as in Norwegian waters and to the south-west of the British Isles. *T. septendrionalis* is a more northern species occurring in Arctic waters and the northern North Sea. There are many other species which may be distinguished by the form of glandular structures on the parapodia.

Travisiopsis lanceolata (**24**.6)

This polychaete belongs to the Typhloscolecidae. The body is some 20-30mm long, transparent when the animal is alive but rather milky after preservation; there are no eyes. There is a T-shaped dorsal papilla anteriorly and this is almost surrounded by a pair of rather elongated nuchal organs. A North Atlantic species occurring to the south-west of Ireland and to the north and west of the British Isles.

Autolytoid stolons

Apart from the holoplanktonic forms mentioned above, the stolons of syllids belonging to the family Autolytinae may be found in the plankton. Autolytoids have a life cycle consisting of a benthonic stock which feeds on hydroids; the stock then buds off stolons posteriorly which have a secondary head and a tail region. Stolons may be budded off singly or in groups but eventually break loose as single sexual phases which are incapable of feeding. Any one stock buds off either males or females. The male stolons have muscular parapodia and branched antennae; the females have simple head appendages. Reference should be made to Hammond (1967) for species likely to occur in the plankton of European waters.

Proceraea (= *Autolytus*) *picta* (**24**.2)

The dorsal cirri are filiform and the anterior ones are dark brown in colour with a fine colourless tip. The body is also dark brown except for a longitudinal mid-dorsal band and two transverse bands anteriorly which are unpigmented. The length is 10-25mm. An Atlantic and Mediterranean coastal species occuring also in the English Channel and North Sea.

Proceraea cornuta (**24**.3)

The colour of this species is variable but there are never longitudinal or transverse white lines. The forked antennae of the male are moderately long. The length reaches 10mm.

Proceraea prismatica (**24**.11)

The horns and the antennae of the male are very long and often coiled and the antennae have transverse rows of glandular structures.

Autolytus edwardsi (**24**.1)

This is a small species normally reaching only some 30mm in length but sometimes up to 10mm and characterized by whitish coloured granules in the dorsal surface of the gut which thus appears as a longitudinal stripe. An Atlantic species occurring in coastal waters to the west of the British Isles in the English Channel and North Sea.

Phylum PHORONIDA

Phoronids are small, gregarious, worm-like animals with a U-shaped gut, inhabitating tubes attached to stones and so on. The mouth is surrounded by a horseshoe-shaped lophophore bearing ciliated feeding tentacles. There is a blood vascular system with red blood corpuscles, a true coelom and a pair of metanephridia. Most are hermaphrodite and the fertilized egg develops into a planktotrophic larva called an actinotrocha. A fully formed actinotrocha is shown in **54**.8. After a cataclysmic metamorphosis this settles on the bottom to form an adult. Phoronids are widely distributed and are found in British plankton, although not commonly. The younger stages are sufficiently similar to the one illustrated to be immediately recognizable. For details of Phoronid development see Forneris (1957).

Sub-Phylum or Class CRUSTACEA

Crustacea, either as adults or as larvae or as both, are the most numerous and conspicuous members of British coarse and medium net plankton and, indeed, may account for over 90% by weight of the zooplankton in many sea areas. Of this great preponderance, by far the most numerous crustaceans are the copepods, with cladocerans running a poor second in most samples. In some inshore waters, however, they (and the copepods) may be outnumbered by the mysids, whilst in some deeper waters euphausids may predominate by sheer bulk. Ostracods, cumaceans, amphipods, and even isopods, may, on occasions, bulk largely in plankton particularly in inshore and estuarine waters, but the general rule holds that copepods are the dominant members of the holoplanktonic Crustacea. With this in mind, the character of the class as a whole can be passed over as being too well known to need repetition and attention can be focussed on those groups of Crustacea important in the holoplankton.

Sub-Class BRANCHIOPODA

Order DIPLOSTRACA

Sub-order CLADOCERA

These small branchiopod crustaceans are distinguished by a bivalved carapace without a hinge which is fused to two or more of the thoracic segments but leaves the head free. There are only four to six trunk limbs, which are of the phyllopod type. There is a single, compound eye. A dorsal cavity below the carapace serves as a brood pouch in which the eggs are incubated. The antenna (second antenna), although it has few joints, is large, bears plumose bristles and is the main locomotory appendage. Most cladocerans live in fresh water. Only two genera are commonly found in British seas and both belong to the group Gymnomera, in which the carapace is much reduced, leaving not only the head but also the limbs uncovered - being, in fact, represented mainly by the brood pouch. *Penilia avirostris* may be found occasionally, but this member of the Calyptomera is so distinctive that it will be readily recognizable.

Podon

Represented by three species, *P. leuckarti*, *P. intermedius* and *P. polyphemoides*, it is easily distinguished from the second genus, *Evadne*, since the head is demarcated from the body by a deep transverse groove. Separation of the species depends on the number of bristles on the exopodites of the trunk limbs and on the shape and size of the body (**25**.1-3). Common in British waters, particularly in spring and summer.

Evadne

Two British species, *E. nordmanni* and *E. spinifera*. There is no groove between head and body. *E. nordmanni* has an oval body bearing a small terminal spine. *E. spinifera* has a more pointed posterior end with a large posterior spine (**25**.4-6). These are mainly warm-water species and are most abundant from March to October.

Sub-class OSTRACODA

The Ostracods are small bivalved crustaceans whose shell length is rarely more than a few mm and in which the second antenna forms a powerful swimming organ. They may be divided into two orders, the Myodocopa in which a heart is present, and the Podocopa which lack a heart. Most of the ostracods likely to be met with belong to the Myodocopa and those found in British waters belong to two main sub-orders, the Cypridiniformes and the Halocypriformes. Reference should be made to Poulsen (1969) for a detailed consideration of the Ostracods likely to be found in the plankton.

Sub-order CYPRIDINIFORMES

The shell is characterized by a curved dorsal border. There is a pair of stalked compound eyes as well as a median nauplius eye below which is a cylindrical or sometimes pointed, frontal organ. The copulatory limbs of the male are paired and none of the limbs are pediform. The fifth limb has a flattened respiratory epidopite bearing up to 60 setae and the seventh limb is characteristic and forms an elongated cleaning organ which curves upwards and posteriorly within the shell and which bears bristles and teeth. The furca is flattened and elongated, the posterior margin only bearing spines.

Gigantocypris mulleri (**25**.14)

The shell of this species is transparent and rounded in outline. It may be distinguished by its large size which reaches 20mm in length in the male but rather less in the female. The frontal organ is very small and the seventh cleaning limb bears at least 150 bristles proximally plus 70 comb teeth on each side near the tip of the limb, paired lateral eyes are very small. A deep-water species occurring to the south and west of Ireland as well as northwards towards the Faeroes and Shetlands. It also occurs occasionally in the Bristol Channel and Irish Sea.

Macrocypridina castanea (**25**.10)

This is a much smaller species than *Gigantocypris* and the shell is never more than 7mm in length. The shell is dark brown except for a transparent region over the eyes. This species occurs to the south-west of Ireland and northwards to the Shetlands but not in the Irish Sea.

Philomedes lilljeborgi (**25**.20)

There is a long tapering frontal organ and the shell is rather oval with a small rounded process on the postero-ventral side of each shell. The overall size is up to 3mm. Adults are common in the plankton from April to June in the north-eastern Atlantic and northern North Sea. It does not occur in the southern North Sea and English Channel.

Philomedes globosa (**25**.9)

The shell is rounded and reaches 3mm in length. It can be distinguished from *P. lilljeborgi* by the presence of numerous shallow pits in the surface of the shell and by the absence of a rounded process on the postero-ventral shell margin. Occurs in the summer months in the north-eastern Atlantic.

Philomedes macandrei (**25**.23)

A small species reaching only 2mm in length and occurring to the north and west of the British Isles. The shell is smooth and characterized by a prominent horn-shaped spike on each side of the rostrum. The ventral part of the hind margin bears a broad process with a toothed border. This species is an Atlantic form which does not occur in the North Sea or English Channel.

Euphilomedes interpuncta (**25**.19)

This is a very small species, the length being approximately 1.5mm. The shell surface is reticulated and is characterized by a pair of spines on the upper and lower corners of the posterior margin. It occurs to the south and west of Ireland, to the north of the British Isles and in the northern North Sea.

Parasterope muelleri (**25**.15)

The shell is nearly circular and has a prominent spiny ridge bearing bristles inside the posterior margin of the shell. The rostrum is bent downwards and there is a fairly marked incisure below it. A north-eastern Atlantic form occurring principally to the west and north-west of the British Isles.

Sub-order HALOCYPRIFORMES

The shell is oval with a straight, or even concave, dorsal margin. The last three pairs of thoracic limbs are pediform, the last pair being reduced and bearing two long, unequal setae. The male copulatory limb is unpaired. There are no lateral eyes; only the unpaired simple median nauplius eye is present. The furca is leaf-like with a rounded posterior border and beset with spines; one big spine projects from the anterior border. There are usually two or more compound glands on the margin of the shell. The glands may be asymmetrically placed on each side of the animal; in many species of *Conchoecia* the right gland opens postero-ventrally and the left gland postero-dorsally.

Euconchoecia chierchiae (**25**.18)

The shell is smooth and thin and bears a compound gland on the posterior margin. There is an obvious spine on the postero-dorsal margin of each shell. A surface-dwelling Atlantic form reaching 1-1.5mm in length.

Halocypris brevirostris (**25**.12)

A small Ostracod with a nearly circular shell reaching a maximum of 2mm in length and without either a prominent rostrum or incisure. The shell bears obvious compound marginal glands; the frontal organ is long and wide and deflected ventrally. An Atlantic form found both at the surface and in deeper waters to the north and west of the British Isles.

Conchoecia elegans (**25**.8)

The dorsal margin of the shell is straight and horizontal and the length reaches 1.5-2.5mm. Asymmetrical glands are present on the shell the right valve of which has the postero-dorsal corner produced into a spine with a pair of minute dorsal denticles. A widely distributed North Atlantic species found throughout the North Sea, to the west and south-west of Ireland and to the north of Scotland.

Conchoecia borealis (**25**.17)

The shell of this species is characterized by the presence of 3-4 teeth on the postero-dorsal margin. Asymmetrical glands are present on the shell valves which reach 3.5mm in length. A widely distributed species occurring in the northern North Sea, as well as to the west and north-west of the British Isles.

Conchoecia obtusata (**25**.13)

The shell of this species has a very convex rather than a truncated posterior margin. This is the smallest of the *Conchoecia* species and the shell length rarely exceeds 1.5-2mm; the surface of the shell is etched with fine concentric grooves. A northern North Atlantic and sub-Arctic species.

Conchoecia haddoni (**25**.22)

The hind part of the shell has a slightly S-shaped border and is devoid of spines. There are longitudinal striations on the antero-ventral part of the shell. The main seta on the first antenna of the male bears in the middle third, 30-47 pairs of small spines. An Atlantic species reaching 3mm in length.

Conchoecia imbricata (**25**.7)

The rostrum is long and curved downwards. Both the antero-ventral and postero-ventral margins of the shell are distinctly toothed. The shell surface in the dorsal half has longitudinal striations but in the ventral half reticulate striations. There are postero-ventral spines as well as postero-dorsal ones and the overall shell length reaches 3.5mm. The main seta of the first antenna has 8-10 pairs of strong, slightly curved spines.

Conchoecia daphnoides (**25**.11)

The rostrum of the female is distinctly longer than in the male and the right shell valve is shorter than the left. The right asymmetrical shell gland opens on the antero-ventral margin, unlike most of the species mentioned above where it opens postero-ventrally. The shell is reticulated all over its surface and the borders are strongly toothed on each side of the pointed posterior margin. The main seta of the first antenna of the male is furnished with a double row of long, curved setae in the first half of its length. An Atlantic species reaching 3-6mm in length.

Conchoecia lophura (**25**.16)

The shell has a series of tightly packed columnar gland cells along the ventral margin. The posterior margin of the shell is sharply truncated and asymmetrical glands are present. An Atlantic species occurring to the west and south-west of the British Isles.

Conchoecia mollis (**25**.21)

The thin shell is some 3mm in length and has a downwardly curved rostrum. There is a characteristic oblique striation on the antero-ventral surface of the shell. A warm-water species occurring in the Atlantic to the south-west and west of the British Isles.

Sub-class or Order COPEPODA

These 'entomostracan' crustaceans are characterized by being of small size; by having a body divisible into head, thorax (bearing biramous appendages) and abdomen (devoid of appendages); by having the head and thorax merging smoothly to form a main or fore-body; by the absence of a shell fold or carapace; and by having a simple, median or nauplius eye with three ocelli (although extra eyes occur in some species). Beyond this it is difficult to make generalizations that would apply to the various sub-orders which comprise the Copepoda.

On embryological and theoretical grounds the body of a copepod is usually regarded as being built of a head, comprising six segments (the first of which has no appendage); a thorax of six segments; and a urosome or abdomen of four segments plus a telson or anal segment which bears the caudal rami or furca (furcal rami). Rarely, however, are the thorax and even the abdomen free from some fusion of segments, whilst those of the head are never apparent externally, although they are indicated by the appendages which are borne on the head. The degree of fusion between segments is variable not only between the main sub-orders (fig. 14) but

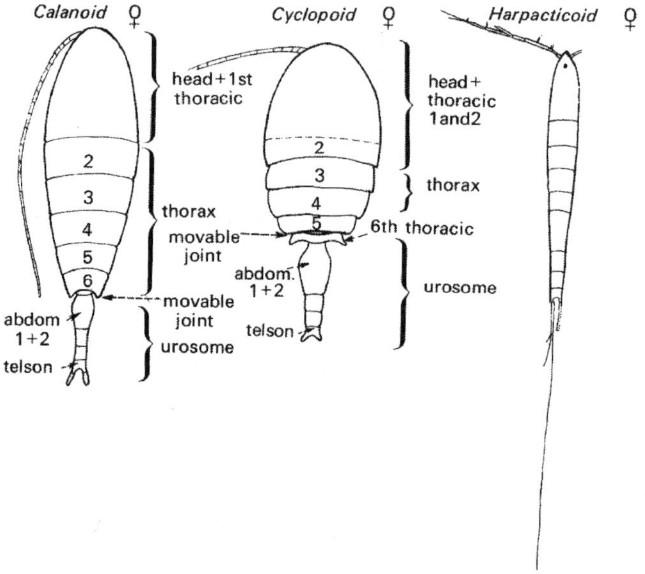

Figure 14. Planktonic copepods - differences in the composition of the body.

within each sub-order, and the external appearance often provides an important clue in identifying various genera and species. Moreover, the segment which bears the genital aperture (and which on theoretical grounds by some is regarded as the last, i.e. seventh thoracic segment) is indistinguishably fused with the first abdominal segment in the female, whilst in the male these two segments remain separate. Finally it may be remarked that there is no agreed system for the enumeration of copepod segments nor for the names of the appendages which they bear. An assessment of the number of segments in each main region of the body is mainly of theoretical interest and will not be further discussed. It may, however, assist in the use of works of reference if the various sets of names of the appendages are given. These are set out in tabular form (below). In the account of copepods which shortly follows, the terms in column 1 are preferred, but those in column 2 are more commonly used in many authoritative works. The third set of terms is, fortunately, rarely met with and has little to commend it.

	1	**2**	**3**
head	antennule	1st antenna	antenna
	antenna	2nd antenna	antennule
	mandible	mandible	mandible
	maxilla	maxilla	1st maxilla (maxillule)
	1st maxilliped	1st maxilliped	2nd maxilla (maxilla)
thorax	2nd maxilliped	2nd maxilliped	maxilliped
	1st swimming leg	1st swimming leg	1st swimming leg
	2nd swimming leg	2nd swimming leg	2nd swimming leg
	3rd swimming leg	3rd swimming leg	3rd swimming leg
	4th swimming leg	4th swimming leg	4th swimming leg
	5th swimming leg	5th swimming leg	5th swimming leg

Sub-order CALANOIDA

Calanoids are by far the most numerous, in bulk and in species, of all truly planktonic copepods in British waters, or, for that matter, in most other sea areas. Their main characters are that the head, together with the thorax, forms a compact, usually ovoid fore-body *clearly distinct from the abdomen*; the line of demarcation (and movable joint) between the two regions lies *behind* the sixth thoracic segment. Only one thoracic segment has fused with the head, and typically the abdomen consists of *four* segments plus the telson and furcal rami. The abdomen is *devoid* of appendages. The antennules (first antennae) are long and are composed of rather numerous segments or joints. The antennae (second antennae) are short. In one group of calanoids (Amphascandria), e.g. *Calanus* and its allies, the antennules, although of different pattern in males and females, are symmetrical. In a second group (Heteroanthrandria), e.g. *Centropages*, one antennule of the male (usually the right) is modified for grasping the female during pairing, the distal portion then being flexed on the proximal part. In a third but small group (Isokerandria), e.g. *Stephos*, the antennules are symmetrical and of the same type in both sexes.

The eggs, except in a few genera, are carried by the female in a single cluster - not in paired egg sacs - and always the eggs are shed freely into the sea. The genital apertures (paired in the female and unpaired in the male) are borne on the first segment of what is usually thought to be the first abdominal segment, but is by some regarded as the last (7th) of the thoracic segments. Certain it is that these apertures appear on the first segment behind the fore-body. As already stated, this segment is fused to the one behind it in the females, so that the rest of the abdomen is in this sex composed at most of three segments plus the telson. The males of many species have one (either the left or the right) of the last pair of thoracic limbs modified as forceps in transferring the spermatophore to the female.

Notes on some of the commoner British members of the calanoids follow and together with the figures should provide sufficient information for a provisional identification. A key to some of the species of copepods illustrated is given in Appendix I p.148. It must, however, be emphasized that an accurate specific diagnosis is a matter for a specialist and will usually involve microdissection and examination of the appendages.

Calanus finmarchicus

One of the principal foods of the herring and one of our largest calanoids. It is particularly abundant in the northern N. Sea and around Scotland, but is rarer well offshore. In more southerly waters it is to a greater or lesser extent replaced by *C. helgolandicus*, which may or may not be a separate species.

C. finmarchicus is separated from *C. helgolandicus* by the nature of the basipodite of the fifth swimming leg. *C. finmarchicus* has a row of teeth along the inner margin of the first basipod segment which follows a slightly convex or only slightly concave

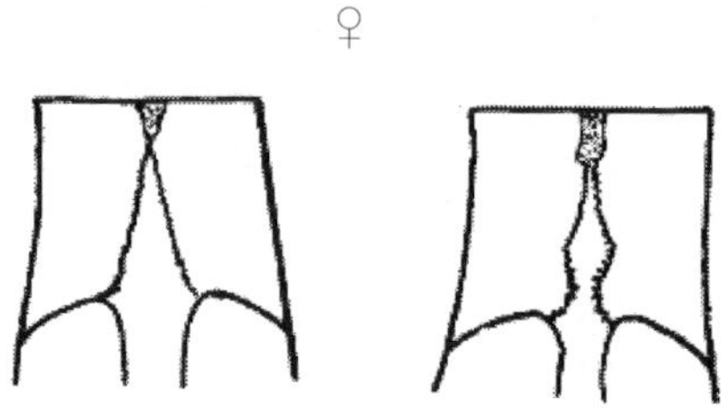

Figure 15. *C. finmarchicus* and *C. helgolandicus* (basal joints)

line. In the females of *C. helgolandicus* this line is strongly concave (fig. 15), In most other respects the two forms are identical. Both species can reach a length of 4mm.

The generic characters most useful in rapid spotting are: the size and general form of the body (**26**.1); five distinct thoracic segments; symmetrical setae or bristles on the furcal rami; antennules longer than the whole body with a large basal segment and usually seen trailing back parallel with the body; and very long plumose setae on the borders of the twenty-third and twenty-fourth joints of the antennae. Apart from *C. finmarchicus* and *C. helgolandicus*, *C. hyperboreus* is typical of colder waters and is distinguished by its larger size, greater transparency, its more obviously plumose setae on the furcal rami and its more pointed head. When met with in more temperate waters it tends to occur in the deeper layers.

Eucalanus elongatus (**26**.2)

The first apparent thoracic segment is fused with the head, and the fifth thoracic segment is very small. The furcal rami are asymmetrical. The head is somewhat triangular in shape.

Rhincalanus nasutus (**26**.3)

There are only four apparent thoracic segments, of which the last is rudimentary. Spines on the thorax and abdomen. Antennules much longer than the body. A pointed head. The left fifth thoracic limb of the male has a very long curved terminal seta. About half the total length of *Calanus finmarchicus*. Mainly a warm-water species, but also present in northern waters.

Paracalanus parvus (**26**.4)

A small calanoid rather less than 1mm. in length. There are only three apparent thoracic segments because the first has fused with the head and the fourth and fifth have fused. No plumose setae on the furcal rami. Widely distributed.

Pseudocalanus elongatus (**26**.5)

Three apparent thoracic segments. Antennules shorter than the body. Rounded front to the head. Fifth thoracic appendage absent in the female, asymmetrical in the male. Rather less than 1.5mm. long. Found in the Channel, but more typical of cold waters.

Microcalanus pusillus (**26**.6)

Three apparent thoracic segments. Antennules only slightly longer than the body. Rather a dumpy fore-body. Fifth thoracic appendage in the male is twice as long on the left side as on the right. About 0.75mm. long. Typical of N. Sea and colder waters.

Euchaeta (**26**.7)

Members of this genus are distinguished by being large; by having four obvious thoracic segments and a sharply pointed head; by the absence of a fifth thoracic appendage in the female and by the presence of large and complicated ones in the male; by long setae on the antennules; by an asymmetrical genital segment; and by having exceptionally long setae on the medial side of the furcal rami.

E. hebes is found mainly in warmer water but has in recent years again extended its range northwards so that it occurs in the western approaches to the Channel.

E. norvegica in this country seems to be confined to lochs on the west of Scotland. This species is sometimes put in a different genus, *Paraeuchaeta*.

Temora longicornis (**27**.1)

Four obvious thoracic segments. Twenty-five joints in the long antennules which are usually seen at right angles to the body. Fore-body rather pear-shaped. Furcal rami long but bearing only short setae. About 1.5mm. long. A very common species in southern N. Sea and most coastal waters.

Eurytemora hirundoides (**27**.2)

Five obvious thoracic segments, the last being incised in the middle so that its edges project sharply to the rear. Antennules shorter than the body, with twenty-four joints. Body about 1mm. long. Sometimes abundant in estuaries and remarkable in being able to withstand a considerable reduction in oxygen tension of the water (XXV. 2).

Eurytemora velox

Characteristic of brackish water and similar to *E. hirundoides*. Distinguished by the rows of small spines along the outer border of the last thoracic segment.

Eurytemora affinis

A brackish water species similar in structure to *E. hirundoides*. The furca is 5-7 times longer than wide, whereas in *E. hirundoides* it is 8-12 times longer than wide.

Metridia lucens (**27**.3)

This species, common around most of our coasts, but characteristic of offshore waters, is distinguished by being of fairly large size (up to 3mm.) and by being one of the few bioluminescent copepods; by having four obvious thoracic segments; and by having antennules about as long as the body. The body, and particularly the urosome, are elongated. Other species of *Metridia* are taken more rarely in British waters.

Centropages

Two species of this genus are fairly common in British plankton. The salient features are that there are five obvious thoracic segments; the urosome has two segments plus the telson in the female but four in the male. The antennules are rather straight and have twenty-two segments or joints. Among the most obvious features of most species are the strong, slightly asymmetrical spines on the fifth thoracic segment. The forceps formed by the right fifth thoracic limb of the male are particularly large, while the setae on the furcal rami are also asymmetrical.

C. typicus (**27**.4)

This species is up to 2mm. long and is common in the Channel and N. Sea. The spines on the fifth thoracic segment are only slightly asymmetrical. It is a euryhaline species.

C. hamatus (**27**.5)

This has a more pronounced asymmetry, particularly in the female. Usually found in waters of lowered salinity. Up to 1.5mm. long.

Isias clavipes (**27**.6)

This species resembles *Centropages* but is distinguished from it by its more rounded fore-body and its smaller size, as well as by its longer and more slender urosome. It reaches a length of about 1.3mm.

Candacia armata (**28**.1)

The corners of the last thoracic segment are pointed or spinous; four obvious thoracic segments; head concave at the sides and squarish in front; urosome slender; the last thoracic segment asymmetrical in males but only slightly so; antennules only slightly shorter than the body; a prominent bristle on each side of the last thoracic segment. Length about 2.5mm. The vast majority of species of *Candacia* can be recognized at once as belonging to this genus by the presence of dark brown pigment on the postero-dorsal region of the fore-body and on the distal segments of the swimming legs. Common in Channel and southern N. Sea.

Anomalocera patersoni (**28**.2)

The head is somewhat triangular with a recurved spine or cephalic hook on each side; it bears two pairs of dorsal ocular lenses. There is a ventral eye and five obvious thoracic segments, the last of which is asymmetrical in the males. The antennule is about half the length of the body and has about twenty joints. It is of very uneven thickness in the male. This is a large copepod and of a blue colour (green after preservation in formalin) as are many other species of the family Pontellidae. Not uncommon in the N. Sea, Atlantic and Channel.

Labidocera wollastoni (**28**.3)

Somewhat similar in general shape to *Anomalocera* but with longer antennules:

and a shorter, more globular urosome composed of only two joints plus a urosome. One pair of dorsal ocular lenses; eye ventral. Fairly common in the Channel and N. Sea.

Parapontella brevicornis (**28**.4)

Four obvious thoracic segments, the last having borders only moderately curved posteriorly and the surface often having numerous black dots. The antennules reach back only as far as the second obvious thoracic segment. Common in the Channel, N. Sea, particularly in coastal waters, and may occur elsewhere.

Acartia

Several species are likely to be met with. The main generic characters are that the body is elongated; there are four obvious thoracic segments; and the antennules are somewhat moniliform with setae of very unequal lengths. Most species are less than 1.5mm long. They are difficult to identify with certainty.

A. clausi (**28**.6)

Common in the N. Sea and Channel. It is euryhaline. Other species are *A. discaudata* (**28**.5) and *A. bifilosa* (**28**.7). Perhaps all three species are commonest in inshore and estuarine waters. *A. longiremis* is one of the commonest copepods in the northern N. Sea.

Very common in water over the edge of the continental slope is *Pleuromamma* the several species of which can easily be recognized by a small black swelling on the side of the cephalothorax.

Of the remaining sub-orders of Copepoda as given by Rose (1933), namely Harpacticoida, Cyclopoida, Notodelphyoida, Monstrilloida, Caligoida and Lernaeoida, only members of the first two are often met with in the plankton, although from time to time monstrilloids may be common in coastal waters. Harpacticoida are mainly bottom-dwelling species; Cyclopoida contains parasitic species, whilst all the members of the remaining four sub-orders are parasites, although they have planktonic stages in their life-histories. It should be remembered, however, that the six sub-orders of copepods other than the Calanoida have often been grouped together as the Podoploea on the grounds that the last true thoracic segment has been incorporated in the urosome, which therefore bears on its first segment a pair of appendages, although these are usually rudimentary.

Sub-order CYCLOPOIDA

As in Calanoids, the fore-body is sharply marked off from the urosome. But always the first, and *sometimes* the second, thoracic segment is fused with the head, whilst the last (sixth) thoracic segment contributes to the urosome whose first segment thus bears a rudimentary pair of limbs. There are at most only four obvious thoracic segments, and usually only three, in the fore-body, and the movable joint between the two main regions of the body is between the fifth and sixth thoracic segments (fig. 14). The urosome consists of five segments in the female and six in the male plus a telson, but some unite. The antennules are short and have only a few joints. The egg sacs are paired in most species.

Oithona

Several species of *Oithona* occur in coastal waters and particularly in those with reduced salinity. Few reach a length of 1mm. Females, recognizable immediately by large paired egg sacs, are often far commoner than males. Four obvious thoracic segments in addition to the last (sixth), which contributes to the urosome and bears bristles representing vestiges of appendages.

O. helgolandica (**29**.1)

Common in northern waters, the Channel and N. Sea, but also reaches much farther south. The females have a pointed rostrum but that of the male is rounded.

O. nana

More typical of warmer waters but widely distributed. It is a smaller species than the previous one, even the females being not much more than 0.5mm.

O. plumifera (**29**.2)

Also widely distributed and immediately recognizable by its plumose antennae and furcal setae.

Cyclopina longicornis (**29**.6)

Widely distributed in shallow water. Only females are known. Other species of *Cyclopina* may occur in the plankton, but all species are mainly benthonic.

Oncaea (**29**.3)

Several species of *Oncaea* are found in British coastal waters as well as those of closely related genera. They are usually orange in colour.

O. venusta (**29**.4)

One of the commoner species. It is a fairly typical cyclopoid but is distinguished by the females having five segments in the fore- and hind-body - three little segments between the telson and the genital segment; six bristles on the furcal rami; and only about six joints in the antennule.

Corycaeus anglicus (**29**.5)

This is the commonest British species of the widely distributed genus *Corycaeus*, a cyclopoid distinguished by having only three obvious segments in the fore-body, and by widely spaced lenticular bodies. The last segment of the fore-body has recurved spinous margins; the urosome has only two obvious segments in the female.

Sub-order HARPACTICOIDA

Although by far the vast majority of harpacticoids are benthonic, some are truly planktonic and, particularly in shallow seas with a sandy or muddy floor, many harpacticoids get swept up so as to be taken in plankton samples. Their most obvious features are their minute size (most being less than 1mm. long) and the lack of obvious divisions between the main regions of the body. Egg sacs may be single or paired and the antennules are short, usually with less than six joints.

Microsetella norvegica (**29**.10)

The body is fusiform and the caudal rami are approximately as long as they are broad. The first segment of the thorax is fused with the head. The setae of the caudal rami are nearly as long as the body but in *M. rosea* they are at least twice the length of the body. *M. norvegica* is a widespread pelagic species occurring in temperate and tropical seas.

Euterpina acutifrons (**29**.7)

In this species the first thoracic segment is fused with the head. The caudal rami are shorter than the last segment and the caudal setae are equal to the last four body segments in length. The head is somewhat pointed anteriorly. A widespread species except in polar and sub-polar seas.

Clytemnestra scutellata (**29**.11)

This species has large epimeral plates on the cephalothoracic and thoracic segments. The first thoracic segment is fused with the head. The caudal rami are shorter than the last segment and the bristles very short. A widespread species in tropical and temperate seas.

Parathalestris croni (**29**.9)

The caudal rami are nearly four times as long as they are broad and are much longer than the last segment. The first thoracic segment is fused with the head. A northern North Atlantic and Arctic species.

Miracia efferata (**29**.8)

This harpacticoid is immediately distinguishable by the presence of cuticular lenses at the anterior end. A widespread temperate and tropical species.

Sub-class MALACOSTRACA

This huge sub-class includes all the larger and more obvious crustaceans, as well as many smaller and less well-known ones. So great is the diversity of its groups that only a brief statement of the recognition features of the sub-class will be given, details being left until the orders and their sub-divisions are considered.

In short, malacostracans are distinguishable by having a rigid pattern of segmentation: six segments in the head, eight in the thorax and six in the abdomen (except in the Leptostraca, which have seven). The thoracic limbs differ sharply in structure from those of the abdomen and from one to three pairs may assist in the feeding, when they are termed maxillipeds. Most malacostracans have the following features: a carapace over the thorax, compound eyes (often borne on stalks), a scale-like exopodite on the antenna (second antenna), biramous antennules, an abdomen which can be flexed under the thorax and a tail-fan borne on the last (sixth) abdominal segment. Most groups of malacostracans have their representatives in the plankton either as adults, as larvae or both.

Super-order PERACARIDA

Malacostraca which incubate the eggs in a ventral brood-pouch formed by plates (oostegites) borne on the endopodites of the thoracic limbs. The carapace, if present, is not fused to the last four thoracic segments.

Order MYSIDACEA

Shrimp-like Peracarida with a shield-like carapace, which covers, but does not fuse with, most or all of the thorax. Stalked eyes are present in all except where they have become vestigial. Swimming thoracic limbs with many jointed exopodites. Brood-pouch formed of up to seven oostegites on the thoracic limbs, one pair of which is modified as maxillipeds. Most of the mysids likely to be met with have an obvious statocyst in each endopodite of the last abdominal appendages (uropods), but some deep-water forms belonging to the families Lophogastridae (**31**.9) and Eucopiidae (**31**.10) have none. Both species illustrated occur to the west and south-west of Ireland and north of Scotland. Most mysids live near, or on, the bottom and are particularly abundant, but seasonal, in inshore and estuarine waters. They migrate to the surface at night and many are euryhaline. Not common in surface hauls except in shallow, turbulent waters.

Siriella armata (**30**.1)

A long, slender mysid with a long, pointed rostrum but a small carapace, which leaves the last three thoracic segments uncovered. A long and slender antennary peduncle more than half the length of the carapace. A lengthy parallel-sided antennal scale whose outer margin is slightly concave and inner margin setose and slightly convex. The outer margin ends in a strong tooth. Eyes on long cylindrical stalks. Telson long, slightly tapering and with an entire margin. Length about 21mm. Widely distributed.

Siriella clausi (**30**.2a-c)

Similar to the above but differs in having a shorter rostral spine, shorter carapace and shorter, thicker eyes. More characteristic of inshore and estuarine waters but widely distributed. Length about 10mm, i.e. about half that of *S. armata.* Other species of *Siriella* occur in British waters, sometimes commonly.

Gastrosaccus sanctus (**30**.3)

A long, slender mysid with a laterally flattened cephalothorax which tapers slightly at the front end. Abdomen cylindrical at the front but flattened from side to side at the level of its fifth segment. A short carapace with a short, blunt rostrum between the bases of the eye-stalks. There is a deep hollow in the mid-dorsal surface of the carapace whose posterior margin is so emarginated as to show the last two thoracic segments on the dorsal side. The posterior margin of the carapace has a lobe projecting forwards on each side. The telson narrows distally and is shorter than the last abdominal segment. It bears six spines on its lateral margin of which the last is the longest.

Gastrosaccus normani

This species closely resembles the previous one but is slightly smaller and usually the posterior margin of the carapace lacks any lobes. A more constant difference between the two species is that the second pleopod is more definitely S-shaped in *G. normani*, which is mainly an offshore species, whereas *G. sanctus* is inshore and confined to waters south of the north shore of the Thames estuary. It lives near the bottom and is commonest in December.

Gastrosaccus spinifer

This is similar to the two previous species, from which it can be distinguished by the more deeply emarginated posterior border to the carapace which bears eight fringe-like projections and a small notch on each side. This species is often found living near the bottom in brackish waters, for it is euryhaline and tolerant of wide temperature changes.

Anchialina agilis (**30**.4)

The body is short and stout and the cuticle is covered with minute bristles. The carapace is large and covers not only all the thoracic segments but also the first abdominal one. It has a straight posterior margin and a large triangular rostrum. This species, as its name implies, is a rapid swimmer and is caught at varying depths. It is mainly a coastal species with a summer maximum.

Leptomysis gracilis (**30**.5)

A slender mysid whose carapace has a pointed rostrum and a concave posterior border, which leaves the last three thoracic segments visible in dorsal view. Its anterior margin has a sharp point on each side. The cuticle is covered by minute oblong scales. This species is common near the bottom in inshore waters and, being euryhaline and eurythermic, often swarms in estuaries. Other species occur in British waters.

Hemimysis lamornae (**30**.6)

A rather short and small mysid. The carapace has a short triangular rostrum and the emarginated posterior border leaves the last two thoracic segments exposed dorsally. A bright orange or red mysid common only locally.

Paramysis arenosa (**30**.7)

A small robust mysid with a carapace bearing a short blunt rostrum and a concave posterior margin. The eyes are short and thick. There is a long apex on the antennal scale. A coastal species swarming among algae on our southern shores.

Schistomysis spiritus (**30**.8)

A slender mysid whose carapace has a rounded, blunt anterior margin and a slightly emarginated posterior one, leaving the whole of the last and part of the seventh thoracic segment exposed in dorsal view. The abdomen is long. There is a long narrow antennal scale. The general appearance of the animal is long and slender and of a transparent nature, with long cylindrical eyes reaching beyond the lateral margins of the carapace. Very common in coastal waters. Other species may be encountered.

Praunus flexuosus (**30**.9)

A long, slender mysid with a narrow carapace with pointed anterolateral regions and a rounded rostrum. The emarginated posterior margin of the carapace leaves part of the seventh and the whole of the last thoracic segment visible in dorsal view. There is a long and stout abdomen sharply curved downwards in the region of the second to fourth segments and upwards beyond this. This is perhaps the commonest mysid around our coasts and one of the largest. It occurs in rock-pools and over sandy grounds or around *Zostera* beds; euryhaline and eurythermic, often swarming in estuaries.

Praunus neglectus

This is very similar to the previous species in structure and habits but has a different pattern of chromatophores which, leaving details on one side, gives it a bright green colour.

P. inermis

This is a more robust species than the two preceding ones but is somewhat similar. A more northern species.

Mesopodopsis slabberi (**30**.10)

A long, slender mysid with a narrow cephalothorax and a short carapace which leaves the last two thoracic segments exposed. The anterior margin of the carapace is rounded, there is a prominent spine on the antero-lateral margins and the posterior margin is nearly straight. The peduncle of the antennule is very long and the antennal scale long and slender. The eye-stalks are also exceptionally long-perhaps the most obvious feature of this species. Exopodites of the uropods long and slender, a third longer then the endopodites. Telson short. It is euryhaline and flourishes in brackish waters. Probably the commonest mysid of the Thames estuary.

Neomysis integer (**31**.1a–d)

The carapace bears anteriorly a triangular plate and a spine on each antero-lateral border. The antennal scale is long and narrow.

Eye stalks are about one and a half times longer than broad. The telson is long, narrow and triangular with twenty to twenty-four spines along its margins. A fully euryhaline species rarely, if ever, found in waters with salinity above 18‰.

Order AMPHIPODA

Peracaridans without a carapace and with a body flattened from side to side; sessile, compound eyes. One thoracic segment is welded to the head and its appendages form maxillipedes, the remaining seven segments forming a mesosome ('thorax'). The six abdominal segments are in two groups of three, forming a metasome and a urosome (which bears the telson) so that the typical malacostracan segmentation of 6; 8; 6; becomes 6+1; 7; 3+3. Each of the seven mesosomatic segments has a pair of coxal plates (probably the equivalents of expanded limb bases). Antennules with two, or in gammarids three, flagella. Antennae uniramous, lacking the exopodite. Thoracic limbs uniramous, the first two of the mesosome being sub-chelate. All bear simple gills (epipodites) and in the female the third to fifth bear oostegites forming a brood-pouch. The third and fourth pairs of thoracic limbs point forward and assist the sub-chelae in food collection, whilst the last three pairs point backwards and are used for crawling. All of the abdominal legs are biramous, the first three pairs forming pleopods and the last three, jumping legs.

Most amphipods are benthonic, but one family, the Hyperiidae, sometimes rated as a sub-order, the Hyperiidea, are fully planktonic, although in shallow-sea areas many bottom-dwelling amphipods may appear in plankton collections - particularly in samples taken at night. Among these, species of *Apherusa* are often common. Even some of the Hyperiidea show a tendency to appear in contact with, or near, other planktonic organisms.

Family HYPERIIDAE

Short-bodied amphipods, usually with large eyes and with the head distinctly marked off from the mesosome. Coxal plates small or absent. Few hyperids are found in British waters. Perhaps the most widely met with is *Hyperia galba*, often a commensal beneath the bells of certain Scyphozoa and hence not often seen in plankton hauls, although it does occur freely in plankton (**31**.2a-c). *H. medusarum*, a more northern species, is distinguished from the previous one by having more setose gnathopods (**31**.3) and by being slightly shorter (about 8-20mm).

Much confusion has existed in the past over the nomenclature of another common genus of pelagic amphipods, viz *Parathemisto*, and before any firm identification of species of this genus is made the paper by Bowman (1960) should be consulted. This also gives a valuable account of the distribution of the various species. In brief, Bowman divides the genus into two sub-genera, *Parathemisto* and *Euthemisto*, and gives the characters of these and of the constituent species. *Parathemisto* (*Parathemisto*) is characterized by (1) the absence of dorsal spines, (2) the flagellum of the first antenna of the female being slender and straight, (3) periopods 5-7 being sub-equal in length, (4) by having the carpus of periopods 3-4 only slightly expanded, (5) by having the peduncle of the third uropod strongly produced at its inner distal angle and (6) the inner ramus of the uropod being serrated on its medial and lateral margins. In contrast, *Parathemisto* (*Euthemisto*) sub-genus has the following characters: (1) dorsal spines present or absent, (2) the flagellum of the first antenna of the female is curved and stouter than in (*Parathemisto*), (3) the carpus of periopods 3-4 are more expanded than in (*Parathemisto*), (4) periopod 5 is much longer than periopods 6-7 and (5) the peduncle of the third uropod is only slightly produced at its inner distal angle.

[Miss J. E. Kane, of the National Institute of Oceanography, tells me, nevertheless, it is very difficult to assign *P. gaudichaudii* to either of the sub-genera. G.E.N.]

In waters around the British Isles only one species of the first sub-genus *Parathemisto* (*Parathemisto*) *abyssorum* will be encountered. As its name implies, it is mainly in deep-water species, although taken in surface hauls in Arctic waters. It occurs only in deeper waters in lower latitudes, and is rare in British waters, except off the east coast.

Parathemisto (*Euthemisto*) *gaudichaudii* (7-25mm. long) can usually be distinguished from *P.* (*Euthemisto*) *gracilipes* (4-7mm. long) by its larger size. Although the ranges of the two species overlap, *P. gaudichaudii* is abundant only in colder, northern water, whereas the reverse is true for *P. gracilipes*. Indeed, *P. gaudichaudii* does not extend as far south as Plymouth today. On morphological grounds *P. gaudichaudii* can be distinguished from *P. gracilipes* by the nature of the third uropod. That of *P. gracilipes* has an inner ramus which is pectinate on both margins, with the inner margin of the peduncle finely toothed in the male but generally smooth in the female. The third uropod of *P. gaudichaudii* always has a smooth inner margin of both the peduncle and inner ramus (**31**.8). The situation is complicated by the fact that *P. gaudichaudii* exists in a long-legged variety (*P. gaudichaudii* var *bispinosa*) and a short-legged form (*P. gaudichaudii* var *compressa*). These are illustrated in figs **31**.4 and 6. See also Dunbar (1963) for Atlantic Hyperiids.

Order ISOPODA

These are peracaridans without a carapace and with a dorso-ventrally flattened body; the compound eyes are sessile. The antennae have a minute exopodite or none at all. There is one pair of maxillipeds and the remaining seven pairs of thoracic limbs are without exopodites and hence are uniramous. Pleopods are respiratory.

Although few isopods are holoplanktonic many bottom dwelling species are taken in the plankton especially in shallow water and amongst floating weed. The groups most frequently encountered belong to two main sub-orders, the Flabellifera, and the Valvifera.

Sub-order FLABELLIFERA

Members of this sub-order have lateral uropods which together with the telson form a tail fan. Two main families are likely to be encountered in the plankton, the Gnathidae and the Cirolanidae.

Family GNATHIDAE

Both males and females as well as the juveniles are of different form but all have only five pairs of walking legs, the first pair having become modified to form gnathopods and the last pair lost. Distinction of species can be made only between males. Two main genera are commonly found, *Paragnathia* in which the gnathopod of the male has five segments, and *Gnathia* in which the gnathopod of the male has two or three segments only.

Gnathia maxillaris (**32**.13)

The front of the head of the male is concave with a slight median projection. Length 4.5-5mm. Widely distributed around the British Isles.

G. oxyuraea (**32**.8)

The anterior margin of the head is tridentate and there is a ridge over each eye. Length 2.5-5.5mm. Widely distributed around the British Isles.

G. dentata (**32**.7)

The anterior margin of the head is tridentate and the mandible bears a large lateral spine. Length 3-4mm. A western English Channel and Irish Sea form.

G. abyssorum (**32**.6)

The anterior margin of the head bears a deep notch. The sides of the head are serrated and the eyes are prominent. Length 2.5-3.5mm. An Atlantic form extending from S.W. Ireland towards the Faroe-Shetland area.

Paragnathia formica (**32**.9)

The gnathopod of the male has five segments. Length 4mm. A widely distributed estuarine species.

Family CIROLANIDAE

The body is laterally compressed with a convex dorsal surface and a flattened ventral surface.

Eurydice gimaldii (**33**.9)

This is a widely distributed offshore species reaching 9mm. in length. The telson bears a pair of postero-lateral teeth and the posterior border is furnished with large plumose setae.

Sub-order VALVIFERA

The uropods are folded ventrally and cover the pleopods.

Family IDOTEIDAE

The two most common genera are *Idotea* in which the abdomen is formed of thirteen segments, the last three of which are fused to the telson, and *Synisoma* in which the abdomen is of one piece.

Idotea balthica (**32**.11)

The sides of the telson taper to a tridentate posterior border. Length up to 30mm. A widely distributed inshore species.

I. emarginata (**32**.10)

The sides of the telson are convex and end in a concave posterior border. Length up to 30mm. A widely distributed inshore species.

I. metallica (**32**.14)

A prominent curved groove behind the eyes. The sides and posterior border of the telson are straight. Length up to 30mm. A western species found only occasionally at Plymouth and not recorded in the N. Sea.

I. linearis (**32**.12)

This is immediately distinguishable by the slender body. Length up to 30mm. A widely distributed inshore species.

Synisoma lancifer (**33**.8)

The telson has concave sides and a pointed, tooth-like posterior margin. An intertidal Channel form.

S. acuminatum (**33**.7)

The body is very narrow and the telson narrows evenly to a point. An intertidal Channel form extending into the Irish Sea.

Order CUMACEA

Peracaridans with a carapace which covers only three or four thoracic segments but which bulges out in front to cover the expanded end of the exopodite of the first thoracic limb. Compound eyes, when present, are sessile. Antenna without an exopodite. First three pairs of thoracic limbs modified to form maxillipeds. Slender uropods. In the female, the second antenna is rudimentary and there are no pleopods. In the male, the second antenna is large and in all families except the Nannastacidae pleopods are present. Four common cumaceans are illustrated in plate **33**.10-13.

Cumaceans are bottom-dwellers which may, on occasions, get whirled up and so collected with true plankton. For their identification see Jones (1957).

Super-order EUCARIDA

Malacostracans whose carapace is fused with all the thoracic segments; stalked, compound eyes, but no brood pouch, the eggs usually being attached to the pleopods.

Order EUPHAUSIACEA

Pelagic shrimp-like crustaceans at first sight somewhat like mysids in having biramous thoracic limbs, but they are easily distinguished from mysids by their larger size and by the fusion of their carapace with all the thoracic segments. All eight thoracic limbs are complete and biramous, none being modified as maxillipeds, nor for prehension. Most species have complex photogenic organs on the bases of some of the thoracic limbs and on the ventral sides of the first four abdominal segments. Only a few genera and species occur in British waters. Being mainly grazers on the phytoplankton, they tend to occur in regions of upwelling where diatoms are most prolific. Millport plankton is often rich in euphausids. Glover (1952) mentions that the euphausids of the N.E. Atlantic fall into three groups:

(1) oceanic species in deep-water areas - *Euphausia krohni, Nematoscelis megalops*; (2) oceanic species over a wide range of depths - *Meganyctiphanes norvegica, Thysanoëssa longicaudata*; (3) shallow-water species - *Thysanoëssa raschii, T. inermis, Nyctiphanes couchi.* Members of groups 1 and 2 occur irregularly and in small numbers in the N. Sea. Group 3 are regular inhabitants of the N. Sea and *N. couchi* and *T. inermis* are indicators of mixed coastal and oceanic water.

The brief notes which follow will help in the identification of euphausids most likely to be found in British plankton.

Nyctiphanes

The peduncle of the antenna is cylindrical and its first segment bears a backwardly directed leaflet. The endopodite of the seventh leg is present but has only two joints. The female has no exopodites on the sixth and seventh legs. *N. couchi* (**32**.5) is a small euphausid reaching only about 17mm in length. Its antennular leaflet is small, pointed, with the sharp end above the inner margin of the antennule. At the end of the second antennular segment on the dorsal side is a four-pointed leaflet or scale.

Meganyctiphanes

In this genus only the eighth leg is rudimentary and the sixth and seventh legs have endopodites in both sexes. *M. norvegica* (**32**.1) is a large species reaching 40mm in length. The front end of the carapace has a convex curve but no rostral spines. A prominent pair of spines occurs on each latero-frontal border and a smaller one about halfway along each lateral margin of the carapace. The first antennular segment has a dorsal backwardly directed leaflet.

Thysanoëssa

The rostrum has a powerful forwardly directed spine. The eye is not spherical, being narrower dorsally. The first six legs have a full complement of joints. The second pair of legs are often much longer than any of the others (this condition applies particularly to pre-adult stages). The seventh leg has a fully formed exopodite.

T. inermis (**32**.3) may reach a length of 32mm. Its rostrum is long and lanceolate, reaching as far forward as the end of the first antennular segment. Eyes are sub-spherical, slightly higher than broad. The last abdominal segment is much shorter than the combined length of the two preceding ones and bears a large spine near its distal end.

T. raschii (**32**.2). The carapace bears small denticles in front of the middle of its lateral margins. The eyes are nearly spherical. The second pair of legs is never very long. The last abdominal segment is much shorter than the combined length of the two preceding ones, but bears no spine.

T. longicaudata (**32**.4). This is a small, slender species never more than 16mm long. The sharp-pointed rostrum reaches beyond the middle of the first antennular segment. There are no lateral denticles on the carapace. The second legs are long and stout. The last abdominal segment is equal to (or only slightly shorter than) the combined lengths of the two preceding ones.

Order DECAPODA

The decapods are distinguished by the fact that the first three pairs of the eight thoracic appendages are modified to form maxillipeds whereas in the euphausids all the eight limbs are unmodified. In the tribe Penaeidea the third legs are chelate and the pleura of the second abdominal segment do not overlap those of the first segment. These two features distinguish the Penaeidea from members of the tribe Caridea to which many of the planktonic decapods belong. In the Caridea the third legs are not chelate and the pleura of the second abdominal segment do overlap those of the first segment. Most of the planktonic decapods are too active to be taken in any but rather large nets or in high-speed samplers and are therefore considered only briefly here. Keys to many of the species are given by Rice (1967).

Pasiphaea multidentata (**33**.6)

The body reaches 100mm in length and has very slender chelae with serrated cutting edges. The mandible has no palp and the telson is strongly notched. The rostrum is represented by a short spine on the antero-dorsal margin of the carapace. A North Atlantic species which occurs in cold sub-Arctic waters as well as to the north and west of the British Isles and in the Bay of Biscay. It tends to be near the surface at night but is found down to 2000 metres during the day.

Pasiphaea tarda (**33**.2)

The body exceeds 200mm in length and as in *P. multidentata* the chelae are slender with serrated cutting edges. It can be distinguished by its large size and the fact that the base of the second leg has less than 5 spines on it whereas in *P. multidentata* the base of this leg has 7-12 spines on it. *P. tarda* is a deep-water species occurring from 250-2500 metres depth in the North Sea, to the north and west of the British Isles as well as in the Bay of Biscay.

Caridion gordoni (**33**.3)

A small species reaching only 25-30mm in length. The cutting edges of the chelae are not serrated and the carpus of the second leg is divided into sub-segments. There are obvious chelae on the first pair of pereiopods. The rostrum is moderately long and reaches beyond the end of the peduncle of the antennule. There is no spine on the fourth abdominal segment, a feature which distinguishes this species from other members of the genus. *C. gordoni* occurs at 50-500 metres depth in the North Sea, English Channel and to the west and north-west of the British Isles as well as in the Bay of Biscay.

Acanthephyra purpurea (**33**.4)

The body reaches 150mm in length. The edges of the chelae are not serrated and the carpus of the second leg is not sub-divided as in *Caridion*. The sixth abdominal segment has a dorsal ridge but there are no ridges on the carapace. The telson bears four pairs of dorso-lateral spines and the eyes are large. *A. purpurea* occurs at 200-600 metres depth in the English Channel and Irish Sea as well as to the north and north-west of the British Isles, to the west of Ireland and in the Bay of Biscay.

Systellaspis debilis (**33**.1)

This species reaches 80mm in length. The rostrum is very long and exceeds the length of the carapace. The posterior margins of the fourth and fifth abdominal segments are toothed and the telson terminates in a point which bears lateral spines. Unlike *Acanthephyra purpurea*, there is no dorsal ridge on the sixth abdominal segment. *S. debilis* occurs from 20-3000 metres to the north and north-west of the British Isles, west of Ireland, and in the Bay of Biscay as well as in the western approaches to the English Channel.

Plesionika martia (**33**.5)

The length reaches some 70mm. The rostrum is very long and characterized by a serrated ventral margin but a smooth dorsal margin except for a few large spines near the base. Found from 150-2000 metres to the west of the British Isles.

Phylum MOLLUSCA

Class GASTROPODA

Sub-class PROSOBRANCHIA

Family HYDROBIIDAE

Hydrobia ulvae (**35**.6)

This species spends a proportion of its time afloat upside-down on the surface film by means of a mucous raft. Length 0.5mm.–3mm., colour black or brown. Found at high water in the inshore plankton of estuaries and in water over mud-flats. Eggs cemented to the shell.

Family JANTHINIDAE

There are three species likely to be found in British waters. All three are permanently pelagic and only rarely come to our S.W. coastal waters. The animals float upside-down on the surface film by means of a bubble raft and the eggs are suspended from an egg raft which is trailed behind. The shell is purple.

Janthina janthina (= *Janthina britannica*) (**35**.9)

Terminal portions of the shell are thick and obvious. The shell is purple and very thin.

Janthina globosa (**35**.5)

The terminal whorls are compressed and insignificant. There is a characteristic longitudinal ridge inside the shell aperture. Purplish in colour.

Janthina exigua (**35**.12)

A very much smaller species. Purplish in colour with a deep groove running around the first whorl.

'HETEROPODS'

This is a group of permanently pelagic gastropods further adapted for this mode of life than the Janthinidae. The shell is greatly reduced and the animal is of a gelatinous consistency. The foot is drawn out into lobes. As in the Janthinidae, most species of these animals float upside-down from the surface film and possess a pair of tentacles with eyes at the base as well as a proboscis containing a radula and bearing a terminal mouth. The eggs are aggregated into gelatinous masses which float on the surface of the water. Only one family of heteropods is found in British waters, the Carinariidae.

Family CARINARIIDAE

Carinaria lamarcki (= *Carinaria mediterranea*) (**35**.4)

This animal is between 25mm and 100mm long. The shell is thin and small with the ctenidium visible beneath the anterior edge. The whole body is semi-transparent. A warm-water species found off S.W. Ireland.

Sub-class OPISTHOBRANCHIATA

The pelagic members of this group are termed 'pteropods'. They are specialized for this mode of life and swim by means of muscular lobes of the foot called parapodia. For the planktonologist the most convenient grouping of the pteropods is into two sub-orders, the Thecosomata and the Gymnosomata. The thecosomes are herbivores, whilst the gymnosomes are carnivores, feeding mainly on the thecosomes.

Sub-order THECOSOMATA

A well-developed external shell is present in all members of the group.

Spiratella (= *Limacina*) *retroversa* (**35**.10)

The smallest of the *Spiratella* species and perhaps the most abundant of all pelagic gastropods. The shell which, as in other thecosomatous pteropods, is sinistrally spiralled, is made up of five whorls and is 2.8mm. long. This species is characteristic of the warm water of the Atlantic and is abundant off Plymouth but is also common in the northern N. Sea.

Spiratella balea (**35**.1)

Length 4.8 mm., found chiefly in water of Atlantic origin, i.e. a similar distribution to that of *S. retroversa*.

Spiratella helicina (**35**.2)

This is a larger species with a lower, less obvious spine.

Euclio pyramidata (**36**.5)

Unlike the coiled shell of *Spiratella*, that of *Euclio pyramidata* is triangular. Length 21mm., i.e. much bigger than *Spiratella*. A Boreal and Arctic species found in waters around the north of the British Isles.

Euclio cuspidata

Smaller than *Euclio pyramidata*, its length being 17mm. Characterized by the presence of long lateral spines on the shell. An Atlantic and Boreal species.

Clio falcata

Length 12.5mm. and no lateral spines on the shell. A Boreal species.

Sub-order GYMNOSOMATA

All the members of this group are active carnivores with an internal shell. The most common species is *Clione limacina*, but *C. gracilis* and *Pneumodermopsis ciliata* and several other species are also found in Atlantic water west of the British Isles (**36**).

Clione limacina (**36**.4)

This is an Arctic and Boreal species with its southern limit at the Bay of Biscay and is found in the N. Sea, and in W. and S.W. England. There is some evidence that the species is composed of northern and southern races. Proboscis and paired head-cones are present as well as lateral wings. The head-cones and tentacles are orange-red, wings pink, liver brown and gonad yellow.

Plymouth specimens are mature at 4-5mm. and do not exceed 12mm. in length whereas Arctic forms reach 41mm. Also, there are more teeth on the radula of the Arctic form (radula formula 14-1-14 or 13-1-13) than in the Plymouth form (4-1-4 or 3-1-3).

Found at Plymouth at any time of the year. Largest specimens in February, March and August. Breeds in the Channel and the numbers do not appear to coincide with influxes of Atlantic water; that is, the presence of *Clione limacina* in the Channel probably does not indicate an influx of Atlantic water as was originally thought.

Clione gracilis

The body is slender, being no wider than the head. Colour whitish. As in *Clione limacina*, there are three pairs of buccal cones and hook sacs with thirty spines or more (Massy, 1939). Radula of eighteen transverse rows, each with a median plate and eight or nine pairs of teeth. Found off S.W. Ireland.

Pneumodermopsis ciliata (**36**.3)

This animal is 12mm. long and has a pale violet tinge. There are cephalic appendages which, apart from the size of the animal, distinguish it from *Clione limacina*; a N. Atlantic species.

Peraclis spp. (**35**.7, 8, 11)

The calcareous shell is coiled into a spiral and the columella is produced anteriorly into a twisted process, the rostrum. The fins are thick and the proboscis short in all species. All are bathypelagic species met with only in very deep hauls to the west and south-west of the British Isles.

Sub-Phylum UROCHORDATA

Class THALIACEA

The Thaliacea is a group of tunicates whose members may be either simple or compound but which never possess a notochord in the adult stage. There is a test below which is a mantle containing muscle bands which are arranged transversely around the body.

Within the mantle is a branchial sac or pharynx which communicates with a peribranchial cavity *via* stigmata which may be few or numerous. The peribranchial cavity leads to the exterior by a posterior atrial aperture. The Thaliacea are subdivided into three orders, the Pyrosomida, the Doliolida and the Salpida. The first order is tropical but the other two have many representatives in British waters.

Order DOLIOLIDA

The doliolids never form permanent colonies and, like all thaliacians, have an alternation of generations between sexual and asexual phases. The sexual phase (gonozooid) is commonly found in the plankton and consists of a barrel-shaped body with an anterior inhalant branchial aperture and a posterior exhalant atrial aperture. The body is encircled by eight hoops of muscle, and the test is extremely thin, the underlying mantle forming a series of lobes around the inhalant and exhalant apertures. The branchial sac occupies the anterior part of the body and communicates with the peribranchial cavity by a series of slits called stigmata.

The eggs derived from this solitary sexual form develop into a tailed larva and this into an asexual oozooid which is termed a nurse or blastozooid. The nurse gives off three types of buds: nutritive gastrozooids, reproductive gonozooids which later become free-living, and foster buds called phorozooids, all of which lodge on a posterior extension called a cadophore. Blastozooids are rarer than gonozooids in our plankton.

There are two main species and one sub-species of *Doliolum* found in British waters and the gonozooids of these can be distinguished fairly easily.

Doliolum gegenbauri var. tritonis (**34**.6)

The smallest sexual forms still bear the stalk of attachment to the asexual oozooid but lose this and soon grow to approximately 17mm. long. Their characteristic feature is that the stigmata start on each side of the mid-dorsal line behind the third muscle band and extend posteriorly to the sixth muscle band (cf. *D. nationalis*). The most anterior stigmata are round and small but they rapidly increase in length posteriorly and join a ventral series which, however, stretches forward only up to halfway between the fourth and fifth muscle band.

A second characteristic feature is that the intestine is very short and wide and curls very sharply ventrally. The ovary is a globular mass in front of the seventh muscle band and opens into the peribranchial cavity. The testis is variable in length but usually long, and crosses muscle band three between the muscle and the ectoderm. It is attached posteriorly to the ovary, and its aperture opens posteriorly into the branchial cavity near the aperture of the ovary. The endostyle starts just behind muscle band two and ends halfway between four and five.

Found in the N. Sea, S. and W. Ireland and Atlantic, and in the Faeroe region.

Doliolum gegenbauri

As might be expected, this species is very similar to the variety *tritonis*. The most obvious difference is that the gonozooid of *D. gegenbauri* never reaches a length of more than 9mm. In addition, the ventral attachment of the gill lamella extends no further anteriorly than muscle band five as opposed to four or four and a half in *D. gegenbauri* var. *tritonis*. Otherwise the structure of the two forms is identical.

It occurs in the northern N. Sea, western English Channel, off S.W. Ireland and off the Faeroes.

Doliolum nationalis (**34**.4)

This species is typically found off S.W. Ireland in the Atlantic but also occurs in the Channel and Southern N. Sea when favourable winds prevail. It may be distinguished from *D. gegenbauri* var. *tritonis* by its smaller size (approximately 3mm.) and by the point of origin of the branchial lamella, for, whilst in *D. gegenbauri* var. *tritonis* the first branchial slit occurs behind muscle band three, that of *D. nationalis* starts behind the second muscle band. Another difference is that the intestine is almost straight and is much narrower. The endostyle starts just behind muscle band two and ends just in front of muscle band four.

Order SALPIDA

The sexual forms (blastozooids) are in chains, whilst the asexual oozooids are solitary and possess a stolon but no gonads. The branchial and atrial apertures are on the anterior and posterior ends respectively and the test is thick. A distinctive feature of the musculature is that it is incomplete ventrally. The alimentary canal may be straight or, more commonly, concentrated with the gonads to form a visceral mass.

Unlike the doliolids, salps do not produce a tailed larva and development is direct. The embryo develops into a solitary oozooid (often termed the solitary form) bearing a stolon. This organ becomes segmented, each segment forming a member of a chain of sexual blastozooids (often termed the aggregated form). The ovary develops before the testis so that self-fertilization cannot occur. There are six species of *Salpa* found in waters adjacent to Britain, all being indicators of warm south-western water. Some species of salps form enormous swarms (e.g. *S. fusiformis*) around our shores.

Salpa (= *Thalia*) *democratica* (**34**.5)

The aggregated forms (blastozooids) have four muscle bands, the first three being fused in the mid-dorsal line. There is only one embryo and the total length of the blastozooid is approximately 15mm.

The solitary form is cylindrical, rather than fusiform as in the blastozooid, and may reach 25mm. in length. There are six muscle bands of which the first is separate, the second, third and fourth fused dorsally, and the fifth and sixth also fused dorsally. A widespread species occurring in the northern N. Sea, English Channel, off S.W. Ireland and off the Faeroes.

Salpa fusiformis (**34**.2)

The aggregated form is approximately 60mm. long but may reach 80mm. The mantle is thick and contains six muscle bands of which the first four and last two are fused. There is a solitary embryo.

The solitary oozooid has a thick mantle and is between 40 and 80mm. in length. There are nine muscle bands as well as the cloacal and buccal siphon musculature. A warm-water species occurring in the northern N. Sea, western English Channel and off S.W. Ireland as well as off the Faeroes.

Salpa (= *Ihlea*) *asymmetrica* (**34**.3)

The aggregated form is oval in shape and between 12 and 25mm. long by 5mm. broad. There are seven muscle bands, the second and third of which lie close together, and the sixth and seventh of which are fused dorsally. There is a quite definite asymmetry in the arrangement of these bands.

The solitary form is 15 to 30mm. long and possesses eleven symmetrical muscle bands. Found in the northern N. Sea, S. and W. Ireland and Atlantic and the Faeroes.

Salpa (= *Iasis*) *zonaria* (**34**.1)

The aggregate form is approximately 35-50mm. long and possesses six separate muscle bands. The solitary form is 30-65mm. long and also possesses six separate muscle bands which, however, are broader than in the other species. Western English Channel, S.W. Ireland and Atlantic and Faeroes. Not recorded from the N. Sea.

Salpa (= *Pegea*) *confederata*

Both the aggregate and the solitary forms of this species are characterized by the extremely short muscle bands which hardly reach the lateral margins of the body. This feature serves to distinguish the species whose solitary form is between 40 and 120mm. long. Found in the western English Channel and S.W. Ireland and Atlantic.

Salpa (=*Thetys*) *vagina*

The aggregate form is normally less than 190mm. long and the muscle bands do not extend ventrally. The solitary form bears two long posterior appendages which increase the length from less than 190mm. to 220mm.; the number of muscle bands is variable but usually large.

Class COPELATA (= Larvacea = Appendicularia)

The appendicularians, as they are usually called, form a group of small pelagic tunicates, some 6-7mm. long, which retain the larval tunicate features of a dorsal notochord in the trunk region and a tail. The animal is surrounded by a transparent 'house' through which water is drawn by movements of the tail. But the 'house' is often damaged or absent in preserved material so that the animal superficially resembles the larva of an ascidian. There is one distinctive feature which can be used in the separation of an ascidian larva and an appendicularian: whilst the tail of an ascidian larva is held in line with the longitudinal axis of the trunk, that of an appendicularian is held at right angles to this axis (**37**.6).

Oikopleura dioica (**37**.7)

This is the commonest British species and often occurs in estuaries. The tail is narrow, being 2-4mm. long and about 0.3-0.6mm. at its widest point. It bears a pair of characteristic large sub-chordal cells about two-thirds along its length on its right side which are readily distinguishable even in unstained material. The ovoid trunk is between 0.5 and 1mm. long and bears either an ovary or a testis posteriorly. Occurs in the northern and southern N. Sea, south and west of Ireland and in the Atlantic, as well as in the Channel, Bristol Channel and Irish Sea.

Oikopleura labradoriensis (**37**.9)

This species is of a similar general appearance to *O. dioica* but can be immediately recognized by the presence of numerous large sub-chordal cells near the tip of the tail on the right side. Unlike *O. dioica,* which is dioecious, *O. labradoriensis* has a single ovary and paired testes in the posterior trunk region. Found in the northern and southern N. Sea, English Channel, S. and W. Ireland and Atlantic, as well as farther north in the Barents Sea and Greenland.

Appendicularia sicula (**37**.4)

The trunk of this species reaches only 0.5mm in length and the tail is small. The fins taper towards the base of the tail and have an obvious notch distally. An Atlantic species occurring to the north west of the British Isles.

Fritillaria borealis (**37**.1-4)

The trunk is long and narrow, being drawn out post-anally to accommodate the testis and ovary. This results in the tail appearing to arise from the mid-ventral surface of the trunk. The tail is short and broad due to the presence of a broad fin. The length of the trunk is approximately 1mm. There are probably two sub-species. *Fritillaria borealis acuta* is a cosmopolitan cold-water form found in polar seas as well as in the southern N. Sea, northern N. Sea, Channel, S.W. Ireland and Atlantic. *Fritillaria borealis truncata* is a warm-water derivative and is found in the northern but not southern N. Sea, Channel, Bristol Channel and Irish Sea, as well as S.W. Ireland and in the Atlantic.

Fritillaria venusta (**37**.2)

The trunk of this species is approximately 10mm long and the tail fins very broad with the posterior margin notched. There is one pair of elongated and rather inconspicuous amphichordal cells. A North Atlantic species occurring to the west of the British Isles as well as in the northern North Sea and Norwegian waters.

Fritillaria tenella (**37**.3)

The trunk of this species reaches 1.2mm in length and there is one pair of rather large amphichordal cells in the tail which has a notched posterior border. A North Atlantic species occurring to the west of the British Isles, in the northern North Sea and Norwegian waters.

Fritillaria pellucida (**37**.5)

There are two pairs of large amphichordal cells and the tail fin is broad. The trunk reaches 2.2mm in length. A North Atlantic species occurring mainly to the west and south-west of the British Isles.

V
General Features of Invertebrate Larvae

MOST groups of marine invertebrates, whether pelagic or benthonic, have larvae which sojourn for a time in the surface waters where they form the bulk of the temporary or meroplankton. Here they feed on planktonic organisms smaller than themselves, grow and either metamorphose into holoplanktonic adults or, if the adults are benthonic, attain a stage at which they leave the surface and sink to the sea-floor. Often the larvae are quite unlike the adults and many have, in the past, been described as separate species. Most marine animals breed in the spring, often over a very short period, but others, e.g. lugworms and some molluscs, are autumn spawners, whilst others again, e.g. *Littorina* sp., have a breeding period extending over several of the warmer months.

It is equally true that practically every major group of animals provides examples which omit a planktonic phase from their life-history. Indeed, in any major group it is possible to arrange the members in a series. At one end are animals that lay large numbers of eggs which hatch into larvae having a protracted planktonic life; at the other end are animals that lay fewer eggs which hatch as post-larval stages so that the planktonic phase is either curtailed or omitted. In between these two extremes, which can be called extreme planktotrophic and extreme lecithotrophic larvae respectively, are all kinds of gradations such as planktotrophic larvae with a short planktonic life, lecithotrophic larvae with a shorter planktonic life, and so on. As the terms planktotrophic and lecithotrophic imply, the variation in the course of development is associated with the amount of food reserves or yolk in the egg. Planktotrophic larvae depend on plankton for food, whilst lecithotrophic larvae are provided with sufficient food to be partially or wholly independent of outside sources for development up to the stage at which they metamorphose. One is reminded of a similar series which can be traced in chordate animals.

Some figures giving the relative abundance of different types of larvae in north temperate waters will be of interest in this connection. Thorson (1946) states that in Danish waters 89% of echinoderms, 70% of polychaetes, 66% of prosobranch molluscs, 92% of tectibranch molluscs and 82% of lamellibranchs - an average of about 80% have planktonic larvae. That is, about 20% omit the planktonic phase and have yolky eggs which may remain enclosed in the maternal oviducts (ovoviviparity), as in several prosobranch molluscs (e.g. *Littorina rudis*), or may become enclosed in hard protective capsules, as in *Nucella* or *Buccinum*, or in strings or cocoons of jelly, as in *Lineus* sp., *Phyllodoce maculata* or *Scoloplos armiger*, many nudibranch molluscs etc., or, alternatively, have some form of brood protection, as in *Spirorbis*, *Harmothoë*, *Henricia* etc. Some of these do not, however, entirely omit a pelagic phase, but the protection afforded by the parent curtails it. Higher percentages of lecithotrophic and non-pelagic development are found in other parts of the world, particularly from Arctic, Antarctic and even deep-sea tropical regions.

In the sections which follow, examples are given of planktotrophic larvae. It can be said of all of them that they are small organisms which, apart from those of crustaceans and tunicates, are ciliated. Usually they have not only powerful locomotory cilia arranged in bands or girdles but also ciliated guts and food-collecting cilia. Their food varies from minute flagellates to larger diatoms and even to larvae of other animals; but the bulk of larvae are herbivorous. On the other hand, larvae themselves are preyed on by planktonic carnivores and clearly a planktonic phase is a hazardous one. For a species to maintain itself at a constant level, on an average at least one pair of larvae must survive to reach sexual maturity. Because of predation and other causes of death, this necessitates the laying of vast numbers of eggs, vast not only absolutely, but even when compared with their nearest land-dwelling relatives. Furthermore, the number of eggs produced by animals having planktotrophic larvae is great when compared with those having lecithotrophic larvae or with those that are viviparous. For example, *Asterias rubens* spawns more than 2,500,000 eggs each season whereas *A. mulleri*, which protects its yolky eggs beneath its disc, lays only about 100 eggs; *Lineus lacteus* lays 100,000,000 eggs, but *L. ruber* lays only a few thousand and encloses them in strings of jelly. Many more examples could be given. If over a period of a year populations of any species remain approximately constant there must be a gigantic wastage of larvae and this is greater than in those species which omit a planktonic phase by resorting to viviparity, brood-pouch protection or extreme yolkiness of the eggs. It may be of interest to consider very briefly the causes of this wastage.

In any large habitat, such as the sea, it is obvious that the chances of a high proportion of the eggs being fertilized is remote if eggs and sperm are shed in a completely haphazard way. The most certain way of ensuring fertilization is by copulation, a solution adopted by nearly all prosobranch molluscs (littorinids, rissoids, stenoglossans etc.), all opisthobranch molluscs (pteropods, nudibranchs), cephalopods and crustaceans. By this method eggs are fertilized in the maternal oviducts and this opens up the possibility of viviparity (as in *Littorina rudis*) or a lesser degree of protection for the eggs. Fertilized eggs are retained for a very variable time and may be quickly released to hatch into planktotrophic larvae, as in *Littorina littorea* and *L. neritoides*, or more yolky ones may be enclosed in jelly (*L. obtusata*), or kept in the oviducts until they are miniature adults (*L. rudis*). But the vast majority of invertebrates shed their eggs freely into the sea though never in a completely haphazard fashion. Usually the chances of fertilization are increased by a simultaneous spawning of males and females and for a great many species it has been shown that the emission of sperms is a stimulus necessary for females to lay eggs. This is true of so many animals that it can be taken as an ecological rule. The chances of fertilization are still further increased by a simultaneous spawning throughout the whole habitat. The classic example is the Pacific palolo worm, *Eunice viridis*, in which the modified hinder ends of the males and females break off and swim in the surface waters preparatory to spawning on two days at the last quarter of the moon in October and November. The synchrony is almost certainly influenced by the duration of the moonlight of the previous full moon, spawning beginning at a fixed interval afterwards. There are many just as good examples from polychaetes in this country; such a one is *Clymenella torquata* which has somewhat recently colonized the north Kent coast from N. America (Newell, 1949b). It spawns on two spring tides nearest to the middle of May (Newell, 1951). A less exact restricted spawning is that of the lug-worm, whose maximum spawning is at the neap tides in October (Newell, 1948). Many other animals, of which the oyster is one, have an extended spawning season, but the oyster has peaks of intensity at certain phases of the moon, though the immediate stimulus to spawning is not known (Korringa, 1957).

Since the number of eggs is greatest in species with planktotrophic larvae, it follows that the greatest wastage occurs in this type of development. Moreover, it is species with this type of life-history which show the greatest fluctuations in numbers of adults, far greater than those with lecithotrophic larvae. Yet it can be shown that the comparative wastage between species which copulate and those whose eggs are fertilized in the sea is not significant. Many tectibranchs, e.g. *Aplysia*, which copulate lay 500,000,000 eggs per annum - far more than many of those which do not copulate.

Predators are undoubtedly the main cause of larval wastage; among them may be listed medusae, ctenophores, chaetognaths, some copepods, various fish (including their larvae), and various filter-feeding invertebrates including benthonic ones. For example, a mussel which filters about one and a half litres per hour may trap and kill about 100,000 lamellibranch larvae in twenty-four hours. Any factors, such as poor food supply, low temperature etc., which tend to slow development and hence prolong larval life increase the chances of predation as well as having direct adverse effects. Currents which tend to take larvae away from grounds suitable for settlement also take their toll of larvae (Evans and Newell, 1957).

The upshot of all this is that a planktotrophic phase is an extremely hazardous one and huge numbers of larvae have to be produced to offset its dangers. For a full discussion see Thorson (1946, 1950). Why, then, it may be asked, is this type of life-history adopted by so many benthonic invertebrates? The usual answer given is that a planktonic larval phase is mainly of use in dispersing the species, thus preventing overcrowding, whilst at the same time offering possibilities of colonizing new territories. This idea should not be accepted uncritically. If the main role of planktonic larvae is to effect dispersion it would be expected that animals with larvae of this kind would be the most widely distributed, but this does not seem to be true. It certainly does not hold good for geographical distribution - rather the reverse seems to be so. For example, as has been mentioned, *Littorina littorea* has a planktotrophic larva whilst its relative, *L. rudis*, which occupies a similar position on the shore, is viviparous. Yet *L. rudis* has colonized both sides of the American continent whilst *L. littorea* is still found only down the east coast. *L. rudis* is the only winkle found on the oceanic island of Rockall. The sea-anemone, *Diadumene luciae*, reproduces only by fission, yet between 1892 and 1896 spread from Connecticut to Plymouth and to Holland by 1913 and the Suez Canal by 1924. *Nereis diversicolor*, extending from the Baltic to Caspian Seas, omits a planktonic phase. So does the lugworm, which is also one of the most widely dispersed of European polychaetes. Of the enteropneusts, only a few genera have pelagic (tornaria) larvae. Enteropneusts, e.g. *Saccoglossus* sp., which omit a planktonic phase, are more widely distributed and commoner than ptychoderids which have a planktonic larva. Examples of this state of affairs could be multiplied over and over again (see Thorson, 1946).

The existence of a planktonic phase allows the adults to be more widely dispersed within the habitat. This may operate favourably in certain species, as, for example, in the ascidians, where the larva, although motile, does not feed. Yet in other species common sense suggests that overcrowding has very little meaning. One has only to consider the dense aggregations on the shore of such animals as barnacles, winkles, limpets or, in offshore waters, of ophiuroids, to realize that many kinds of animals thrive when crowded and some could not maintain their numbers if dispersed. Indeed, it has been shown (Cole and Knight-Jones, 1939, Knight-Jones, 1951, 1953a and b, and Knight-Jones and Crisp, 1953) that some animals, e.g. barnacles, *Spirorbis*, ascidians and probably many others, aggregate when they settle just before metamorphosis, being apparently attracted by chemical stimuli emanating from those which had settled previously. This is the reverse of dispersal, a process which helps to explain the patchiness in distribution of many animals on the sea-floor, for rarely are they scattered at random. Sometimes this patchiness is associated with differences, often subtle, in the nature of the substratum. A good example of this is the little polychaete, *Ophelia bicornis*, which has been studied intensively by D. P. Wilson (see Wilson, 1952, for a summary). *O. bicornis* lives in this country only in the sands around the Exe estuary and is common only in Bullhill Bank.

The yolky eggs hatch out into lecithotrophic larvae which soon become negatively phototactic and prepare to metamorphose after about ten days. They can, like so many larvae, delay metamorphosis until they reach a suitable substratum and only Bullhill Bank sands will evoke a high percentage of metamorphoses. Larvae of the polychaetes, *Owenia fusiformis*, *Notomastus latericeus*, *Scolelepis fuliginosa*, *Pectinaria koreni* and *Melinna cristata*, can all delay metamorphosis until they reach an acceptable substratum (Wilson, 1952; Nyholm, 1950). Veligers of *Teredo* and of *Cardium* behave in a similar manner. From all this it seems that a great variety of animals with pelagic larvae include features in their life-history which act, so to speak, as antidotes to dispersal. Such features include behavioural ones, such as an ability on the part of the larvae to delay metamorphosis until a suitable substratum is encountered, and a response to settle preferentially in areas already occupied by adults of the same species. This tendency to aggregate is not in the least surprising - once it has been observed - for a suitable substratum is often somewhat rare. Submerged or floating timber, into which a lone *Teredo* can burrow, is scarce in the sea. No wonder that most pieces are riddled with *Teredo* burrows!

If, then, a larval phase is not in most species primarily concerned with dispersal, what is its survival value? It lies, perhaps, in providing a phase in which the rich food supplies of the plankton can be exploited. The food requirements of the larvae usually seem to be (weight for weight) greater than and understandably very different from those of the adults. It has been calculated, for example (Zeuthen, 1947), that a 10mg. larva utilizes food at five times the rate of an adult weighing a gram. Moreover, with a relatively shorter gut, it is doubtful if larvae digest and absorb food as efficiently as adults and they therefore require greater concentrations of food. An oyster larva, for example, eats about 1,000,000 nanoplankton organisms before it is ready for metamorphosis (Thorson, 1950). Consequently, the main spawning season of British invertebrates is in the spring and summer and such as spawn in the autumn usually lay large yolked eggs which make them less dependent on the plankton.

VI
Polychaete Larvae

DESPITE their abundance on the sea-floor, particularly in sandy and muddy shallow-sea areas, polychaete larvae do not bulk largely in most plankton samples. The reasons for this are various. Firstly, most polychaetes spawn over a very short period - often on a few tides only. Secondly, those which hatch as trochophores in the strict sense of the term are not common and in preserved samples are difficult to spot. Indeed, many, if not most, polychaetes have a short planktonic phase only and this is easily missed. Yet even in modern textbooks it is usually implied, if not stated, that polychaetes have a trochophore larva which has a lengthy planktonic life, although as long ago as 1911 Shearer pointed out the error of this view (see also Segrove, 1941). Newell (1951) reviewed the evidence as follows:

'The planktonic trochophore phase, far from being the "typical polychaete larva", would seem to be rather unusual and true trochophores are common only amongst members of the Serpulidae and Phyllodocidae. Many polychaetes from other families have pelagic larvae with reduced ciliary girdles and a review of all the types of known polychaete larvae would reveal that a graded series could be recognized between larvae with no prominent ciliary girdles (and usually well supplied with yolk) and the serpulid type of trochophore. This view seems to be endorsed by the findings of Thorson - (1946), who states that only about 70% of the polychaetes of the Oresund whose development is known have pelagic larvae. Not all of these, however, are typical trochophores. About 13.7% live for only a short time in the plankton and many do not feed there to any extent. Others, representing about 6.8%, do not feed on the phytoplankton at all, but depend upon their reserves of yolk before taking to the bottom. About 5.5% of the total number of species whose development is known reproduce in a variable manner and may produce either pelagic or non-pelagic larvae and about 1.4% are viviparous. To put it another way, about 45.2% of the polychaetes whose development was recorded by Thorson always have a long pelagic life but may hatch at a later stage than the trochophore which is a feeding stage hatching from a small yolked egg, whilst about 27.4% of the polychaetes of the Oresund have non-pelagic larvae. One may hazard a guess that further investigation would probably increase the numbers in this group, since Thorson was able to determine the course of development in only 51% of the total number of polychaetes found, and larvae never met with in plankton hauls would be the most likely to escape detection.'

Subsequent studies on polychaete life-histories have confirmed these views.

Nolte (1936 and 1938), Thorson (1946) and Rasmussen (1956) are among the most useful sources for the identification of polychaete larvae and post-larval stages (as well as for references to previous literature), but much information has been added since then, particularly for certain families, e.g. Spionidae, Disomidae and Peocilochaetidae (Hannerz, 1956 and 1961). Unfortunately, it has not been collected together but remains in scientific journals, many of which are not readily available. For example, the extensive studies of Okuda are in *J. Fac. Sci. Hokkaido Imp. Univ.* 2 and the reprints bear no date! Many polychaetes have a very wide geographical distribution and the literature is correspondingly scattered. The difficulties of identifying polychaete larvae and post-larvae found in plankton samples are therefore very great. Moreover, since even closely related species may have a very different type of embryonic development and may produce quite different kinds of larvae, it is difficult to place a larva, even approximately, by extrapolation from species whose development is known. For this reason the classification of larvae on general form, rather than taxonomically, seems to have little value (see, however, Nolte, 1936 and 1938). Nevertheless, post-larval stages can often be referred to families by utilizing features in which these stages resemble the adults, particularly in eyes, prostomial appendages and chaetae.

Since, however, one of the types of larvae, the trochophore, is found in molluscs, and in a slightly different form in other invertebrate phyla, as well as in the polychaetes, it deserves special mention. As described by Shearer (1911) and Segrove (1941), the trochophore is a transparent, top-shaped larva with a capacious blastocoel, a pair of protonephridia, a fully formed functional ciliated gut, and a prominent apical plate; it swims by means of two ciliary girdles, a prototroch and a metatroch (**38**.1). In this typical form a trochophore is practically confined to the serpulids, but to a varying degree most polychaete eggs are charged with yolk and from them hatch larvae in which a blastocoel is either small or absent and in which protonephridia are lacking. The ciliation on larvae of this type is very variable (Segrove, 1941). In a few, for example that of *Clymenella torquata*, the surface is uniformly ciliated. In others, for example ariciids (Anderson, 1961), ciliary girdles are numerous.

Many larvae pass through a stage equivalent to a trochophore whilst enclosed in the egg membrane, and hatch with the rudiments of a number of segments, some of which may bear larval chaetae. Such are termed metatrochophores. By the addition of more segments they become post-larval stages or juveniles, the head and appendages gradually approaching the adult condition. At the same time the larval chaetae may be replaced by ones resembling those of the adult, thus providing valuable clues as to the specific identity of the organism. A distinct larval stage which retains ciliary girdles but has also parapodia and chaetae (often very long ones) is called a nectochaete.

The larvae of a few polychaetes are sufficiently distinctive to be recognized at a glance, a good example being the mitraria larva of *Owenia fusiformis* (Wilson, 1932) (**39**.4). Others may fairly readily be identified down to families and even genera.

It has been found in practice that there are three main ways in which polychaete larvae can be related to the adults. The most certain is to fertilize the eggs, either naturally or artificially, in the laboratory and to examine the resulting larvae. This is possible only in rather a few species. The second is to collect larvae at intervals from the natural habitat and attempt to place them in a series the oldest members of which are sufficiently like the adult for a specific diagnosis to be made. Thirdly, the adult worms are collected at intervals and the state of maturation of the gonads noted. This enables the breeding period to be anticipated. Then if polychaete larvae appear in plankton samples immediately after the adults of a certain species have spawned, the strong presumption is that the larvae belong to the species in question. This method is particularly applicable to worms which have a restricted spawning period - as many of them do. Experience has shown that even comparatively rare worms spawn sufficient eggs for their larvae to be picked up very readily from plankton hauls. Indeed, very often larvae are recorded in areas from which the adult has not been taken. Similar methods are applicable to many other invertebrates but have rarely been used. From this, it is clear that in the identification of any polychaete larva it is vital to pay particular attention to the time of year at which it occurs in the plankton and to compare this with citations in the literature. Examples of planktonic polychaete larvae from some of the commonest families are illustrated in **38** and **39**. That many of the more abundant families have no representatives is in part due to lack of information and in part to the fact that many of them lack a planktonic phase.

VII
Crustacean Larvae

IN COMMON with nearly every major group of animals there are some crustaceans which lay large numbers of eggs containing small amounts of yolk, whilst others lay very yolky eggs. The general rule is that the less the yolk, the younger the stage at which the eggs hatch. Examples from every sub-class of Crustacea can be found in which a rather simple larva called a nauplius hatches from the egg. This is a minute larva with no obvious segmentation. It bears only the first three pairs of head appendages: antennules (uniramous), antennae (which are the main swimming organs), and mandibles, both of which are biramous. There is a simple, median eye, a large labrum and a pair of frontal organs in the form either of papillae or filaments. The commonest nauplii are those of copepods and of acorn barnacles which can be distinguished from the copepods by a triangular carapace prolonged on each side into a frontal horn (**40**.1), whereas those of copepods have an elongate oval carapace ending in a point posteriorly (**40**.5a, b). Yet in many examples from each of the sub-classes a free-swimming nauplius stage is omitted, the larva which hatches from the egg having grown to a later stage of development, as is shown by the possession of more than three pairs of appendages.

Then follows a succession of larval stages in which more and more segments and limbs are added, though not necessarily in the adult order from front to rear. Always each successive stage is attained by a moult, so that sudden changes in form may occur, but in the branchiopods and a few others the adult stage is reached more gradually by the addition of a few segments at a time from a growth zone just in front of the telson. In such, the development is said to be gradual, in contrast to metamorphic. Only those larvae which undergo sudden changes at ecdysis have larval stages sufficiently distinct to receive special names. Gurney (1939) prefers to acknowledge only three major types of larvae: (1) the nauplius, which swims by means of its antennae; (2) the protozoea, swimming like a nauplius but having a partially segmented thorax and long abdomen, with compound eyes still not functional, a carapace not fused to the thorax and a forked telson, and with only maxillipeds 1 and 2 functional as biramous limbs; and (3) the zoea-mysis stage in which the main divisions of the body are obvious but in which the abdomen, now clearly segmented, is at a relatively more advanced stage than the thorax although the animal swims mainly by means of the biramous maxillipeds. Each of these three main types of larvae may undergo a series of moults and, in some crustaceans, such larvae receive special names. Usually, biramous limbs behind the maxillipeds arise without any sudden change in the form of the larva so that the zoea passes gradually into the mysis or schizopod stage which has biramous thoracic limbs behind the maxillipeds. Probably each species has a fixed number of moults in its life-history and for many animals it is now customary td refer to these by serial numbers rather than to give a multiplicity of names. A key to some of the common crustacean larvae is given in Appendix 2.

Class or Sub-class CIRRIPEDIA

All cirripedes (barnacles and the parasitic Rhizocephala) have six naupliar stages which can be distinguished from those of other crustaceans by the fronto-lateral horns on the carapace. In British waters cirripede nauplii occur in two main broods. That in March-April consists mainly of nauplii of *Balanus balanoides* and *Verruca stroemia*; that in July-September of *B. perforatus* in western waters and of *B. improvisus* and *Elminius modestus* in the southern N. Sea. After several moults there is a sudden change to a cypris stage which has a bivalve shell, a compound eye, six pairs of thoracic appendages and a short abdomen. It is this stage which settles and changes into the adult. Rhizocephalan larvae are unlikely to be commonly met with.

The nauplii of various species can be separated by their different sizes and by the nature of the setation of their limbs. Commoner species can also be identified by their general appearance (Bassindale, 1936).

Chthamalus stellatus (**40**.7)

Mainly a barnacle of exposed rocks at high shore levels on western coasts. It has a nauplius with a rather rounded outline and with a rounded distal end to the labrum. The caudal end is very short, and the limb setae are finely plumose. The fifth naupliar stage is approximately 0.5mm long.

Balanus balanoides (**40**.2, 4)

Widely distributed. The nauplii are to be found all round our coasts. Seen from above, the outline of the body is triangular and in later naupliar stages has a pair of posterior spines. The tip of the labrum is truncated. The first stage is about 0.3 mm., reaching 0.9mm. in the fifth stage.

Verruca stroemia (**40**.1)

Widely distributed, but not as common as the other species. The nauplii are triangular in outline. The labrum has a rounded tip. The caudal region is longer than that of *B. balanoides* and the fronto-lateral horns are longer than in either of the two preceding species. The final stage nauplius is approximately 0.6mm. long.

Elminius modestus (**40**.6)

The dominant barnacle in south-east waters, spreading to other regions. The nauplius is similar to that of *B. balanoides* in shape and in having a posterior pair of spines on the carapace. But the labrum is trilobed, the large median lobe projecting posteriorly. The stage two nauplius is approximately 0.4mm. long.

Sub-class MALACOSTRACA

Order EUPHAUSIACEA

Meganyctiphanes norvegica and *Thysanoessa inermis* shed their eggs into the sea but others, e.g. *Nyctiphanes couchi*, retain them beneath the thorax until the contained larvae have reached a more advanced stage of development.

From the eggs of *M. norvegica* and *T. inermis* hatches a nauplius which is soon followed by a second stage nauplius and then, after another moult, by a metanauplius (**40**.3). This is in turn succeeded by three calyptopis stages (corresponding to the protozoea stages of penaeids). The first calyptopis (**41**.5) has three post-mandibular limbs and a segmented thorax but no abdominal segments and is about 0.5mm long in *M. norvegica*. The eyes are covered by the carapace. By an increase in size and the addition of a five-segmented abdomen, the second calyptopis stage is reached and in the third stage the sixth abdominal segment is added. Next follows a furcilia (**41**.1) larva (corresponding to a zoea) in which the eyes project beyond the carapace and in which abdominal limbs (pleopods) are added successively. Breeding is mainly in winter; later larvae in spring.

The furcilia of *M. norvegica* differs from that of *T. inermis* (**41**.4) by having a broad carapace ending anteriorly in a truncated rostrum. The carapace of *T. inermis* is narrow and ends anteriorly in a small, rounded protuberance. Five pleopods are present in *T. inermis* before the first becomes setose, whereas in *M. norvegica* the first pleopod is setose by the time the fourth has appeared.

Order DECAPODA

Sub-order NATANTIA

Tribe PENAEIDEA

These decapods are not common in the inshore plankton and the larvae are likely to be encountered mainly to the west and south-west of the British Isles. Characteristically the telson is deeply notched to form a pair of terminal spines. The first three pairs of legs often bear chelae and there are setose exopods on all thoracic limbs including maxillipeds. The larva of *Gennadas* is shown in **47**.2.

Tribe CARIDEA

Family PANDALIDAE

As in all other decapods except penaeids, a nauplius stage is omitted and the first larva is a protozoea in which the main segments are already delimited and a carapace and compound eyes are present. Pandalid zoeae always have a rostrum which in later stages becomes toothed. The antennal scale (exopodite) is segmented. Of the five thoracic limbs behind the three which form maxillipeds (but which in early larva are mainly swimming organs), all except the fifth have exopodites. The endopodite of the second pair is chelate in later stages. Caridean larvae always have two pairs of chelate walking legs in contrast to penaeids and stenopids which have three pairs. Larvae of two species of pandalids are common in inshore plankton from March–June and may extend into autumn. The first zoea of *Pandalus montagui* (**41**.2) has a full complement of limb rudiments whereas that of *Pandalina brevirostris* (**41**.3) has none. The first zoea of *P. montagui* is about 3-4mm long, about twice the size of that of *P. brevirostris*, and has a much longer rostrum - so much so that in later stages its toothed tip projects in front of the eyes.

There are in all up to nine larval stages, each one being larger than the preceding one and more closely resembling the adult. Those in which the five thoracic limbs are biramous and form swimming organs are termed schizopod, or mysis, larvae. When the thoracic limbs have lost the exopodites to become uniramous the larva is a post-mysid.

Family HIPPOLYTIDAE

Hippolyte varians and *H. prideauxiana* have life-histories and larvae essentially like those of pandalids and their larvae are often seasonally abundant in inshore waters around our coasts; those of *H. varians* are commoner than those of *H. prideauxiana* which does not occur in the North Sea.

H. varians (**41**.8)

This has five larval stages in each of which there is a small lateral spine on each side of the fifth abdominal segment. There are three spines on each side of the carapace. The conspicuous rostrum is pointed and the three pairs of maxillipeds of the first zoea stage form the main swimming organs. The compound eyes are covered by the carapace until the second zoeal stage is reached. First stage zoea 1.25mm. long.

H. prideauxiana

The larvae have five lateral spines on the carapace and those on the fifth abdominal segment are long. First stage zoeae of this species are about 1.6mm. long, that is, like those of *H. varians*, they are rather small.

Spirontocaris spinus (**47**.8)

The rostrum is prominent and there are paired spines on the posterior margins of the fourth and fifth abdominal segments. There is also a transverse row of short setae on the dorsal surface of the fourth abdominal segment. The zoea is 4-8mm in length and is found to the west and south-west of the British Isles as well as in Norwegian waters and in the northern North Sea.

Lysmata seticaudata (**47**.7)

The larva is quite characteristic with the fifth leg (the eighth thoracic limb) being longest with an expanded terminal section (propodus). The eyestalks are also very long. The body is from 2-7mm in length and the antero-ventral margin of the carapace is toothed. This species is likely to be found in the western English Channel and in the Bay of Biscay but not to the west and north of the British Isles.

Caridion gordoni (**43**.10)

The rostrum is nearly as long as the carapace and supra-orbital spines are present in most stages. There are dorso-lateral spines on the posterior margin of the fifth abdominal segment. The zoea is from 3-12mm in length and is found throughout the North Sea as well as to the west and south-west of the British Isles, in the English Channel and Bay of Biscay.

Family OPLOPHORIDAE

Acanthephyra purpurea (**43**.2)

There is a fat-filled dorsal ridge on the third abdominal segment and a pair of lateral spines on the posterior margin of the fifth abdominal segment. The postero-ventral margins of the carapace are serrated and so are the margins of the first two abdominal segments. The length is from 3.5-12mm and the body is transparent with scattered red chromatophores. Found mainly to the west and north west of the British Isles and in the Bay of Biscay.

Systellaspis debilis (**43**.4)

There is no fat-filled dorsal ridge in the third abdominal segment. The body is

10-12mm in length and orange in colour. The thorax is filled with red yolk granules and there are photophores on the side of the carapace and abdomen. This species is found to the west and north-west of the British Isles and in the Bay of Biscay.

Family ALPHEIDAE

Athanas nitescens (**47**.6)

The fifth leg is much longer than the fourth and the endopod of the first maxilliped is small. The first zoeal stage is approximately 1.5mm long and reaches 3mm in the final stage. Found in the North Sea, English Channel, to the west and north-west of the British Isles and in the Bay of Biscay.

Family PROCESSIDAE

Life-histories of members of this family resemble closely those of preceding ones. Moreover, processid larvae can be distinguished only with great difficulty from those of pandalids, the sole clear-cut feature being that the exopodite or scale of the antenna has no sign of segmentation whereas in pandalids and hippolytids it is clearly jointed. The absence of a rostrum in stage one processid larvae (which are about 3mm. long) also serves to separate them from some pandalids, as does the presence of a pair of small dorso-lateral spines on the fifth abdominal segment. Adult processids are rarer than would perhaps be suggested by the abundance of their larvae from spring to autumn in inshore waters. The larvae of *P. canaliculata* (**41**.6) and *P. edulis* are often met with.

Family PALAEMONIDAE

Although palaemonid prawns of several species are common in most areas, their larvae are so rarely encountered in plankton hauls that they will be omitted from this account.

Family CRANGONIDAE

Crangon crangon (**41**.7) [Note: labelled as *C. vulgaris* in Plate 41].

Larvae of this species, the commonest of our shrimps, are especially common in the spring in inshore plankton. Even the first stage zoea is large (2mm.) and the last stage mysis is about 6.5mm. long. The three maxillipeds have natatory exopodites, but the remaining thoracic appendages (legs) do not begin to appear until the third stage larva and are not all present until the fifth stage. Even then the third and fourth legs lack a natatory exopodite. The inner ramus of the antennule is an unjointed spine. The antennal scale is unjointed and there is a large dorsal spine on the posterior margin of the third abdominal segment.

Crangon allmani (**42**.3)

Larvae of this species are larger than those of *C. crangon*, being 2.5mm. long in the first and 6.5mm. in the final stage. The chief points of difference between the larvae of the two species are the absence of a dorsal spine on the posterior margin of the third abdominal segment of *C. allmani* and the enormous size of the lateral spines on the fifth abdominal segment.

Philocheras fasciatus (**42**.1)

The larvae closely resemble those of the two preceding species but differ in having two spines on the posterior margin of the third and fourth abdominal segments. The peduncles of the antennules are proportionally longer and may be twice the length of the rostrum. The animal has a distinctively dark coloration and is much smaller than the corresponding larva of *Crangon crangon* and *C. allmani*, the final stage larva being only 3.5mm. long.

Pontophilus norvegicus (**47**.9)

This zoea reaches 10-16mm in length and has prominent dorso-lateral spines on the fifth abdominal segment. There is a large dorsal spine on the third abdominal segment and the posterior and ventral margins of abdominal segments 1-5 are serrated. The antero-ventral margin of the carapace is also toothed. A widespread species in the North Sea and English Channel as well as in the Irish Sea and to the west of Ireland.

Tribe PALINURA

Only the larvae of the rock lobster, *Palinurus vulgaris*, occur at all commonly in British plankton and then only in western and south-western waters. There is a reduction in the number of early larval stages and from the egg hatches a unique form of mysis stage called a phyllosoma which is found elsewhere only in scyllarids (an example is *S. arcticus* occasionally caught off Plymouth). By its transparency, expanded body and immensely long appendages it is well suited to planktonic life (**42**.4). It is followed by another planktonic phase but one which inhabits the deeper layers. This is a puerulus stage and is rarely taken in plankton hauls.

Tribe ASTACURA

The early larvae of *Homarus vulgaris*, the common lobster, are never common in the plankton except in a few districts.

Nephrops norvegica (**47**.5)

The zoea is quite characteristic with the first three pairs of legs chelate and with setose exopods on all five legs (thoracic limbs 4-8 incl.). The telson is produced into a pair of long serrated spines and there are long dorsal spines on abdominal segments 3 and 4.

Tribe ANOMURA

Larvae of two main anomuran families frequently occur in our plankton, those of the Galatheidae and of the Porcellanidae. The first stage larva which hatches from the egg is a zoea, unique in that the carapace bears a long rostral and *two* long postero-lateral spines. The first zoeal stage has two fully formed maxillipeds but only a rudimentary third one. Only in the later stages does the third maxilliped become a fully natatory appendage. There are five larval stages in galatheids and three in porcellanids, the later ones being comparable to what in brachyuran crabs would be called megalopae.

Family GALATHEIDAE

There are two genera, *Munida* and *Galathea*, whose larvae are commonly found on plankton hauls - particularly those of *M. bamffica*, *G. strigosa*, *G. dispersa* and *G. squamifera*.

M. bamffica (**42**.2)

The first larval stage is about 6mm. long (large compared with those of *Galathea* sp.). The spiny tip to the antennal scale is unique among galatheid larvae. All the abdominal segments bear paired dorsal spines whilst the fourth and fifth have lateral spines in addition. The main season is January–June.

G. strigosa (**43**.7)

The first zoeal stage is 3.5mm. long and the second 5mm. long. As in other *Galathea* sp., the rostral spine and antennal scale are proportionally smaller than in *Munida*. No dorsal spines on the posterior margins of the abdominal segments. The second–fifth abdominal segments are serrated on their posterior margins. The telson bears eight spines in all except the first stage. Uropods, when they appear, have nine posteriorly projecting spines.

G. dispersa (**42**.5)

The first zoeal stage is only 2-5mm. long and the final stage is 6.8mm., as compared with 7.0mm. in *G. strigosa*. Lateral spines are lacking from the fourth abdominal segment and the postero-lateral spines on the carapace are relatively shorter than in the preceding species.

G. squamifera (**42**.6)

The first larval stage is the smallest of the three species mentioned, being only 22mm. long. The fourth or final larva is only 4.8mm. long. The abdomen bears short lateral spines on segments four and five - much shorter than in *G. strigosa*. Other *Galathea* species occur as larvae, but only rarely in British plankton.

Family PORCELLANIDAE

Although larvae of this family share with the Galatheidae the usual anomuran features, there are obvious differences between the larvae of the two families. This is particularly true of the carapace spines which here are greatly exaggerated so that they may be twice, or even three times as long as the body of the animal in the early stage zoea. Unlike *Galathea* larvae, there are no serrations on the carapace and the telson has a convex, instead of a concave, posterior margin. The antennal scale is a mere spine whereas in *Galathea* it is quite broad. The fifth abdominal segment has no pleopod and uropods are lacking even in the late larval stages.

Porcellana platycheles (**43**.1)

This species is distinguished from *P. longicornis* by the relative lengths of the rostral and posterior spines of the carapace. In *P. platycheles* the posterior spines are half the length of the rostral spine. Common in the inshore plankton of both the western and north-eastern British Isles from March to autumn.

P. longicornis (**43**.3)

This is a much more widely distributed species. The posterior spines are only one-third the length of the rostral spine. Common in inshore plankton but also found in offshore water from March to August.

Tribe THALASSINIDEA

Family AXIIDAE

Axius stirhynchus (**43**.8)

The larva has a long rostrum with a toothed margin and dorsal spines on abdominal segments 2-6. There is a pair of antero-lateral spines on the carapace. The telson has a long median spine. The exopodites of the maxillipeds bear four terminal bristles and the thoracic exopodites behind these (except those of the fourth and fifth segments) are setose. Uropods and the first pair of pleopods are not formed until the post-larval stage is reached.

Family LAOMEDIIDAE

Jaxea nocturna (**43**.6)

This is the only British member of the family and the larva is a type of zoea called a trachelifer. It has an unmistakably slender appearance owing to the great length of the region immediately behind the eyes. The rostrum is small and the telson lacks a median spine. There are only two maxillipeds in the first larva and the third develops later. There are six larval stages, each of which is not uncommon in inshore plankton.

Family CALLIANASSIDAE

Callianassa subterranea (**43**.9)

The larvae of this species are fairly abundant in spring to early autumn plankton. After a first stage there are five which correspond to zoeae; each of these has a long serrated rostrum and there is a dorsal serrated ridge on abdominal segments 3-5. A single dorsal spine is found on abdominal segment three and pleopods occur on segments 3-5. The telson has one median and seven lateral spines on each side. The antennal scale is unsegmented in the first larva, in which swimming exopodites are present on the three maxillipeds.

Upogebia deltaura (**43**.11)

The larvae of this species are common in late July and August. The rostrum is much flatter than in *Callianassa* and is not toothed. Also, unlike *Callianassa*, the abdominal segments are not toothed, whilst the telson bears a short median spine.

U. stellata

The larvae occur mainly in May and June and resemble those of *U. deltura* except that they are larger in the earlier stages and the antennae and last abdominal segments are relatively longer.

Tribe PAGURIDEA

Pagurus is the commonest genus containing as it does all the common hermit crabs. There are four larval stages and a post-larval one, once thought to be a separate genus and still termed the *Glaucothoë*. Early larvae resemble those of the galatheids. There are similar projections from the carapace - two at the rear and one at the front - but never any serrations on the lateral margins of the carapace. The endopodite of the uropods is so small as to be called vestigial in pagurid larvae.

There are two families, the Paguridae and the Lithodidae, and their larvae are easily distinguishable.

Family PAGURIDAE

There are four zoeal stages in which the eyes are characteristically longer antero-posteriorly than the width of the abdomen. The uropods are present in the last two zoeal stages and the telson, which is only slightly concave posteriorly, bears six or seven pairs of spines.

The megalopa also has uropods and is not very spiny.

Pagurus bernhardus (**44**.2, 6)

The zoea has a long straight antennal spine which is at least six times as long as broad. The rostrum projects beyond the spine on the antennal scale and the fifth abdominal segment bears a pair of small lateral spines. A characteristic feature is that the longest spine on the telson is more than half the greatest width of the telson. The first zoea is about 3.5mm. and the final stage reaches 8mm.

The megalopae of all species belonging to this genus possess four pairs of functional pleopods. The chelae of those of *P. bernhardus* have a smooth propodus and the antennae reach as far as the tip of the chelae. The eye-stalks are about twice as long as they are broad. The length is approximately 4.2mm.

This is a very widely distributed species found all around our coasts.

Pagurus pubescens (**44**.3)

The zoeae of this species are similar to that of *P. bernhardus* but the fifth abdominal segment has large lateral spines and the rostrum is somewhat shorter, reaching only the tip of the spine on the antennal scale. The first zoea is approximately 3mm. and in the final zoea attains a length of 5.5mm.

The megalopa (**44**.8) is also similar to that of *P. bernhardus* but the eye-stalks are not so long in relation to their breadth, being only one and a half times as long as broad. The right chela is larger than the left. The megalopa is about 3mm. long.

P. pubescens is fairly widely distributed around our coasts, being found in the north and west of the British Isles and in the Atlantic. It is also abundant on the north-east coast of England.

Pagurus cuanensis

Zoeae of this species differ from those of *P. bernhardus* and *P. pubescens* in that the antennal scale is curved and only four times as long as it is broad. The longest telson spine is less than half the greatest width of the telson and the fifth abdominal segment bears prominent lateral spines. Paired yellow chromatophores are present on the carapace in stage one and two zoeae and also on abdominal segments 5 ad 6. The first zoea is approximately 2.7mm. long and the final stage rarely exceeds 4mm. in length.

The megalopa (**44**.9) is characterized by the hairy chelae (which, however, bear few spines); by the spines on the carpus of the second and third legs; and by the antennae which do not reach as far forward as the tips of the chelae. Usually about 3mm. long.

Larvae of *P. cuanensis* are found in spring (April) in the north and west of the British Isles, in the English Channel, and also occur in the southern N. Sea.

Pagurus prideauxi (**44**.5)

The zoeae are similar to those of *P. cuanensis* but the paired yellow chromatophores are present on the fifth abdominal segment only. Additional features are the orange chromatophores on each side of the carapace and the presence of a small mid-dorsal spine on the posterior margin of the sixth abdominal segment. The longest spine on the telson of the first and second zoeal stages is just less than a third the greatest width of the telson. The length is from 3.5 to 6mm., i.e. much larger than those of *P. cuanensis.*

The chelae of the megalopa (**45**.1) have spines on the propodus and the antennae do not reach as far forward as the tip of the chelae which are spiny but only moderately hairy. There are no spines on the carpus of the third legs. The megalopa is about 4.3mm. long.

Larvae of *P. prideauxi* are found from March to April to the north and west of the British Isles in the Atlantic and northern N. Sea, as well as in western parts of the Channel, but, like those of *P. cuanensis*, are not found in the southern N. Sea.

Anapagurus

As in *P. cuanensis* and *P. prideauxi*, the antennal scale in members of the genus *Anapagurus* is relatively shorter and broader than in *P. bernhardus* and *P. pubescens* and is curved. Zoeae of *Anapagurus* sp. resemble those of *P. cuanensis* in that the longest spine on the telson is less than half the greatest width of the telson, but differ in that the fifth abdominal segment bears only short lateral spines (*A. hyndmanni*) or none (*A. laevis*). The final stage zoea has only two (*A. hyndmanni*) or three (*A. laevis*) pairs of pleopods as opposed to four pairs in *Pagurus* sp.

Anapagurus laevis (**44**.1; **45**.2)

The zoea (**44**.1) is characterized by the presence of a chromatophore on the mid-dorsal surface of the carapace and one on each eye-stalk. In addition, the posterior margin of the telson is straight, or even slightly convex, in early stages. The first stage zoea is approximately 3mm. and the final stage 5.5mm. long.

The megalopa (**45**.2) has three pairs of functional pleopods and eye-stalks only slightly longer than broad. It is about 3.5mm. long.

Larvae of this species are found from February to May, but also in July and August in the western English Channel, north and west coasts of the British Isles and in the northern N. Sea, i.e. like those of the other species mentioned, the larvae are possibly found in the southern N. Sea.

Anapagurus hyndmanni (**44**.4; **45**.6)

Zoeae (**44**.4) of this species differ from those of *A. laevis* in the absence of chromatophores on the centre of the carapace. Instead, chromatophores are present on the bases of the maxillipeds and one on the rostrum. The rostrum is appreciably longer than the antennal scale. The first stage zoea is approximately 2.7mm. and reaches 4mm before metamorphosis into the megalopa (**45**.6) which has only two pairs of pleopods and is 2.7mm. long.

Larvae of *A. hyndmanni* are found in the northern N. Sea, English Channel, and on the west and north coasts of the British Isles, and probably in the southern N. Sea.

Family LITHODIDAE

There is only one species occurring in our waters, *Lithodes maja*. There are two zoeal stages (**44**.7) which are characterized by the large number of spines (eight or nine pairs) on the telson, and the concave posterior margin of the telson. The longitudinal diameter of the eye is shorter than in the Paguridae and is less than the width of the abdomen. There are no uropods, and the length of the first zoea is 7.6mm., reaching 7.9mm. in the final (second) stage.

The megalopa (**45**.3) has no uropods and is very spiny. As in the zoea, there are small pleopods on the sixth abdominal segment. Length about 4.5mm.

This is a more northern species found in the northern N. Sea, southern N. Sea, N.W. Scotland, W. Ireland and Atlantic, but not in the English Channel.

Tribe COENOBITIDEA

Family DIOGENIDAE

There are two genera belonging to this family, both of which are of a characteristic appearance.

Diogenes pugilator (**45**.7; **45**.4)

The zoeae (**45**.7) of this species are very small; the first stage is only 1.3mm long and after five stages it reaches only 2.8mm. In the first stage, the telson is invaginated, but in later stages is straight and there are, by this time, two pairs of pleopods.

The megalopa (**45**.4) is readily distinguishable by the very much larger left chela and by the relatively short abdomen which is about the same length as the carapace. The megalopa rarely exceeds 1.6mm. in length. *D. pugilator* is a warm-water species occurring mainly in the Bay of Biscay but also found in the Bristol Channel, Irish Sea and S.W. Scotland, and in the English Channel.

Clibanarius erythrops (**45**.8)

The zoeae are small, the first stage being approximately 1.7mm., and have a broad, rounded rostrum quite different from any of the species previously mentioned.

The megalopa has a rounded telson which is somewhat broader than long and the endopods of the uropods are very short. The megalopa is approximately 2.5mm.

Larvae of *Clibanarius erythrops* are found in the Bay of Biscay and occasionally in the western English Channel.

The most recent accounts of Decapod larvae are those of Macdonald, Pike and Williamson (1957) , Williamson (1957) and Pike and Williamson (1958). These give a comprehensive key for their identification and many figures.

Tribe BRACHURA

A great variety of British crabs have planktonic larvae and only a few of the common and widely distributed species can be mentioned here, but locally - as, for example, over sandy bays - others such as *Portumnus latipes* may be dominant (for a full account see Lebour, 1927, 1928a and b). All hatch as protozoeae (probably the equivalent of a penaeid metanauplius - Gurney, 1926). The protozoea changes into the first zoea stage which is flattened from side to side with a carapace that has a rostral and dorsal spine and, in many species, short lateral spines also. Compound eyes, not as yet on stalks, mouth-parts and natatory first and second maxillipeds, walking legs hidden beneath the carapace, and a long abdomen finally consisting of six segments and ending in a forked spiny telson are other important features.

The number of zoeal stages varies with the species and seems to be two, four or five but never three. It culminates in a megalopa which swims by means of its abdominal appendages (pleopods), the maxillipeds by now functioning as accessory mouth-parts. The body is dorso-ventrally flattened and this stage could be described as having all the essential features of a young crab into which it is transformed by a single moult.

Family DROMIIDAE

Dromia personatus (**45**.5, 9)

The zoeae (**45**.9) of this species are heavily pigmented, especially on the abdominal segments and on the telson. There are no spines on the abdominal segments but the postero-lateral margins of the carapace are prolonged into long spines. There are five stages of which the last stage has four legs. Of these, only the first bears a setose exopod and the other three have rudimentary ones. There are four pairs of pleopods. The first stage is approximately 3mm. and the fifth stage 6.5mm.

The carapace of the megalopa (**45**.5) is about 3.2mm long and the abdomen is somewhat flexed. There are four pairs of pleopods, and uropods are prominent. *D. personatus* is a warm-water species occurring in the Bay of Biscay but its larvae are also occasionally found in the western regions of the English Channel.

Family PORTUNIDAE

Two of the commonest British genera of crabs, *Portunus* and *Carcinus*, belong here. Their larvae bulk largely in most plankton hauls.

Portunus puber (**46**.1, 5)

The zoeae (**46**.1) are often abundant in April and the megalopae reach their peak in July. The first zoea is about 1.8mm. long and has long dorsal and rostral spines on the carapace as well as short lateral ones. The pleopods appear as small knobs in the third zoeal stage at which time lateral spines become apparent on the abdomen. The black and orange carapace bears lateral bristles. In later zoeal stages the flagellum of the antenna grows much longer and the pleopods branch. The final zoeal stage is about 4mm. long and has two, instead of three, spines on the telson by which time the first walking leg has become chelate. The megalopa (**46**.5) stage which follows is somewhat smaller (about 3mm. long) than the last zoea.

P. puber larvae are very similar in the zoeal stages to those of *P. depurator* but are readily separable in the megalopa stage since in the former the rostral carapace spine is bent ventrally whereas in the latter it is in line with the longitudinal axis of the body. In the North Sea the commonest portunid larvae are those of *P. depurator* and *P. holsatus*.

Carcinus maenas (**46**.2, 8)

Each of the four zoeal stages (**46**.2) (of which the first is about 1.4mm and the last 3.2mm. long) can be distinguished from those of *Portunus* by the absence of lateral spines on the carapace and on the abdominal segments. The megalopa (**46**.8) can be separated from that of *P. depurator* by the absence of dorso-lateral knobs on the carapace and from that of *P. puber* by the fact that its rostral spine is in the same straight line as the main axis of the body. Moreover, there are only five bristles on the uropods of the megalopa of *Carcinus* as compared with ten on those of *Portunus*. The zoeae of *Carcinus* have a distinctive dark chromatophore which extends along each side of the carapace.

Family CANCRIDAE

Cancer pagurus (**46**.7)

This is the commonest member of the family and the larvae appear in our plankton from April to August. The zoeal stages closely resemble those of *Portunus*, but when alive can be distinguished by being redder with much less black pigment, having shorter rostral spines, and having an antennal exopodite about one-third of the length of the antennal spine, whereas in *P. puber* it may be as much as half.

The megalopa (**46**.7) is easily distinguished from that of *Portunus* or *Carcinus* by the very pointed rostrum and the prominent dorsal spine projecting to the rear but bending upwards near its tip. Moreover, each of the last pair of pleopods bears eight bristles in *Cancer*, ten in *Portunus*, and five in *Carcinus*.

Family CORYSTIDAE

Corystes cassivelaunus (**47**.3)

The masked crab is an inhabitant of clean sandy shores at low tidal levels and below; it has five zoeal stages. These and the megalopae are often abundant in the inshore plankton from March to June. The zoeae are larger than those of any other crab. Dorsal and rostral spines are long and there are also lateral spines on the carapace. In later stage larvae the flagellum of the antenna is noticeably long, longer in fact than the antennal spine. The first zoea is about 2.4mm. long and at the fifth stage about 7.5mm, whilst in the megalopa (**47**.3) the carapace alone attains a length of 3mm. Other features of distinction are three teeth on the anterior edge of the carapace, one in the middle of its dorsal surface and four on each side of the carapace just behind the eyes. As in the larva of *Portunus*, each of the last pair of pleopods bears ten pairs of bristles.

Family XANTHIDAE

Pilumnus hirtellus (**46**.3, 9)

This species is often abundant and its larvae may occur in vast numbers in inshore plankton. The zoeae (**46**.3) are smaller than those of previously mentioned crabs and have short dorsal and rostral spines. There are three lateral spines on the telson. The megalopa (**46**.9) has no spines on the carapace and its rostrum points downwards. The protozoea is about 1mm. long and metamorphoses in about three hours into the first zoea which is brownish in colour, has rudiments of the walking legs but no pleopods, and lateral spines on the second–fifth abdominal segments. The second zoeal stage is about 1.8mm. long and at this time pleopods appear, whilst lateral spines appear on the first abdominal segments. The third zoeal stage is slightly longer than the second whilst the fourth (about 2.5mm. long) changes into the megalopa which, unlike that of *Corystes* or *Cancer*, has no spines on its carapace although there are several dorsal knobs. The last pair of pleopods bears six bristles each - a distinguishing feature.

Family MAIIDAE

Maia squinado

Larvae of the spiny spider crab occur mainly in late summer plankton. The protozoeae are about 2.5mm long and soon change into the first zoea (**46**.4) stage which has a greenish colour because it has yellow and black chromatophores. Lateral spines are short, and dorsal and rostral spines curved. The first zoeal stage is unusual in having a flagellum to the antenna as well as having pleopods. There are six abdominal segments even in the first zoea. Lateral spines are present on the second to fifth abdominal segments and those of the third and fifth increase in length until the second (last) zoeal stage which has long pleopods but small walking legs. It changes into a megalopa (**47**.1) which is about 2.5mm long with a broad rostrum ending in a thin spine. The last pair of pleopods bear five bristles each and the carapace has numerous protuberances on its dorsal surface.

Inachus dorsettensis (**46**.6, 10)

This is not a very common crab. It has larvae occurring in the late summer and autumn plankton. The zoea (**46**.6) is distinctive in that there is only a dorsal spine on the carapace, rostral and lateral spines being absent. Moreover, there is a large lateral spine on the telson. There are only two zoeal stages, the first being about 2.5mm long and the second 2.9mm. The megalopa (**46**.10) is smaller, only 1.6mm. long. It has two large spines near the base of the rostrum as well as a number of spines on the dorsal surface of the carapace. The walking legs are long and slender (as in the adult) and each of the pleopods bears eight setae.

Family LEUCOSIIDAE

Ebalia tuberosa (**47**.4)

Larvae are not uncommon in the inshore plankton of the western British Isles throughout the year but are most abundant in spring and summer. The zoea is characterized by its small size and the extreme reduction of the lateral spines. Both dorsal and rostral spines are absent.

Zoeas of some deep-water south-western and western decapods are also illustrated in **47**.

VIII
Gastropod Larvae

Sub-class PROSOBRANCHIA

THERE are three main periods in the year during which prosobranch larvae become common in the plankton: spring, summer and autumn, the majority of species occurring in the summer plankton. Roughly 60% of these larvae have a long pelagic phase (Thorson, 1946), about four-fifths of them being derived from epifaunal parents.

Most prosobranchs encase their eggs in gelatinous or horny masses which are attached to the substratum so that eggs of prosobranchs are rarely found in the plankton. Nevertheless, there are a few exceptions to this and they are mentioned below. Descriptions and keys to the identification of many prosobranch larvae are given by Fretter and Pilkington (1970).

Order ARCHAEOGASTROPODA

The eggs of most archaeogastropods are shed singly into the water, e.g. in *Acmaea virginea* and in *Gibbula cineraria*, and the larva hatches as a trochophore. But in many genera the eggs are laid in gelatinous masses which are attached to the substratum and the larva hatches as a veliger, i.e. the trochophore stage is omitted. Typically, the veliger is provided with a *round* ciliated velum.

Order MESOGASTROPODA

This group includes the vast majority of the common intertidal prosobranchs. The eggs of non-viviparous forms are always encased singly or in clumps in a gelatinous material and are normally attached to the substratum. Rarely, as in *Littorina littorea* (**48**.3) and *L. neritoides* (**48**.1), the eggs occur in the plankton. The first free-living stage is a veliger (**48**.2, 4, 5). This is characterized by the possession of a *lobed* velum. The number of lobes varies; it is usually two, or rarely four, in the newly hatched larva but may increase to four or six in later larvae. Whorls are continuously added during larval life and may be as many as eight before metamorphosis.

Some mesogastropods are viviparous, e.g. *Littorina saxatilis* (= *rudis*), and so a pelagic larval stage is absent.

Family LITTORINIDAE

Littorina neritoides (**48**.1, 2)

Egg capsules are present in the inshore plankton during winter and spring (April). The capsules are approximately 0.16mm. across and 0.09mm. high and contain a single egg (**48**.1). The veliger (**48**.2) is shelled and bears spiral striae and tubercles as in some rissoids and is yellowish in colour. The velum is colourless in the early veliger but later has black markings as in *L. littorea.* Not present in the plankton of the southern N. Sea or Channel east of the Isle of Wight.

Littorina littorea (**48**.3, 4)

The eggs (**48**.3) are extremely common in the inshore plankton at Plymouth February to June, at Millport January to July, and at Heligoland and Thames Estuary March to May. Egg capsules are 0.96mm. across, i.e. much larger than those of *L. neritoides.* Each capsule contains up to nine eggs but normally there are only two or three eggs per capsule. The larvae hatch after five days as veligers with one and a half whorled shells and are characterized by the presence of two dark pigmented patches, one on each velar lobe (**48**.4). These show through the shell when the animal has retracted into it. The length of the larval shell before metamorphosis varies but may reach 0.6mm; it is pale yellowish in colour and has no sculpturing (c.f. *L. neritoides*). The larvae are in the plankton for two weeks before settling.

The other common British littorinids, *L. littoralis* (= obtusata) (eggs laid in jelly) and *L. saxatilis* (= *rudis*) (viviparous) have larvae which hatch as the crawling stage.

Family RISSOIDAE

Alvania crassa (**48**.5)

Larvae of *Alvania crassa* are common in the summer and autumn plankton at Plymouth (Lebour, 1935). The newly hatched larva is 0.1mm. with a spiral, sculptured shell which has a pronounced beak. Later whorls are not sculptured. The larva is colourless or yellowish and the shell reaches 0.4mm before metamorphosis. The foot is brown at this stage. Probably not present in the plankton of the northern N. Sea.

Family CERITHIIDAE

The larvae of this family are characterized by their small size, rounded bibbed velum and the presence of a process from the outer lip of the shell. This process projects between the lobes of the velum. Three common species whose larvae are abundant during spring and summer are described below.

Triphora perversa (**49**.4)

Larvae of this species are abundant in the summer at Plymouth but common also in spring in both inshore and offshore plankton. The newly hatched veliger is 0.16mm. long with a sinistral one-whorled shell ornamented with fine raised dots, replaced by striae in later larvae. The shell is an opaque brown colour and the velum colourless. The rest of the animal is yellowish. The velum of the newly hatched larva is 0.18mm. across and formed of two round lobes.

Later, the shell becomes two-whorled, the upper one and a half whorls being dotted and the rest striated. At this stage it is 0.2mm. high. When the shell is 0.64mm. high there are six or seven whorls, the first one and a half being dotted and the rest striated. Finally, the seventh and eighth whorls become tuberculate, the velum is lost and the larva settles. Often one velar lobe is larger than the other.

Cerithiopsis tubercularis (**49**.3)

Larvae common in both inshore and offshore plankton. The youngest larvae are about 0.24mm. and the shell has two and a half whorls. The outer lip of the shell bears a characteristic large plate-like process which protrudes anteriorly and bears a number of concentric striations. The largest larva is 0.64mm. long and the suture lines of the shell are dark. The shell is characteristically smooth and the final whorl shows traces of the adult tubercles. The velar lobes are round, one being larger than the other. The animal is yellowish but the foot acquires dark pigment in the late veliger.

Cerithiopsis barleei (**49**.2)

Larvae common in the late spring and summer plankton. Very similar in structure to *Triphora perversa* except that the shell has purplish-red markings on the columella and sutures. The veliger spends about five weeks in the plankton and settles when the shell has reached the four-and-a-half-whorled stage. Distinguished from *C. tubercularis* by its smooth shell.

Family EULIMIDAE

Balcis alba (= *Eulima alba*) (**49**.7)

The veligers are very common in August–September, off Plymouth. The shell is 0.16mm. across and of one and a half whorls in the early veliger. It is characterized at this stage by its broad apex, which soon, however, becomes pointed. The velum is 0.2mm. across and bibbed.

A very conspicuous feature, even in the retracted larva, is the presence of black pigment around the mouth and beside the oesophagus. The spire lengthens in the later larva, which has a dark three-lobed foot with black lines running towards the mouth. Length of late larva is approximately 0.72mm.

Balcis devians (= *Eulima philippi*) (**50**.6)

Larvae of this species are common between July and September. The veliger is approximately the same size as *B. alba* but the apex to the spire is blunter and there is no black pigment. The digestive gland is yellow-brown. The shell of the late larva reaches 0.64mm. before the velum is lost and the animal settles.

Family NATICIDAE

Natica catena (**48**.6)

The larvae are confined to coastal plankton during the spring and summer. The first larva has a smooth shell of one and a half whorls and the velum is bilobed and colourless. Later, a wide umbilicus develops in the shell and the border of the velum becomes purple or brown. The shell of the late larva is approximately 0.5mm. long and has two and a half whorls. The larva closely resembles that of *Nassarius reticulatus* but the latter has a shell siphon and bears a tooth-like process on the outer lip of the shell (XLV. 6).

Natica alderi (**48**.7)

The larvae of this species are common in both inshore and offshore plankton. The veliger is larger than that of *N. catena*, reaching 0.8mm. in length before metamorphosis and having a shell with three and a half whorls. A dark spot develops at the end of each lobe of the four-lobed velum which becomes very large. The veliger resembles that of *Nassarius incrassatus* (**49**.5) but the latter has a shell siphon and a peg on the outer lip of the shell, a distinguishing feature.

Family CYPRAEIDAE

Trivia monacha (**48**.8)

This species breeds throughout the year but veligers which later develop into an echinospira are most common in the Channel during spring and summer. Common in inshore plankton but rather rarer offshore. The first larva has a bilobed velum, an accessory shell, a small true shell and rudimentary tentacles. The body is full of orange-yellow yolk which later disappears. As the tentacles become longer, the animal becomes darker owing to the development of a brown liver and purple gut. The velum has a brown border with the lateral margin of each lobe slightly concave. The width of the accessory shell of the first larva is 0.36mm. and the true shell 0.16mm.

The anterior lobes become longer than the hinder ones and scattered pigment appears on the sole of the foot which develops lateral lobes. Finally, before metamorphosis, the accessory shell is covered by the mantle and absorbed. Operculum and lateral lobes of the foot disappear and the larva settles.

Family LAMELLARIIDAE

Lamellaria perspicua (**48**.9)

This species spawns from January to May and the larva, called an echinospira, has a coiled shell some 4mm. long. Common in the Atlantic, N. Sea and English Channel.

Velutina velutina (**48**.10)

The larvae of this species are common in spring and summer plankton and measure about 2mm. across. The larva is a typical echinospira but differs from that of *Lamellaria perspicua* in the absence of protuberances on the shell.

Order NEOGASTROPODA (= Stenoglossa)

The majority of the Stenoglossa hatch from the eggs as a crawling stage. There are, however, exceptions to this generalization and *Nassarius* and the turrids have planktonic larval stages. The eggs are usually encased in a horny membrane and are always attached to the substratum.

Family NASSARIIDAE

Nassarius reticulatus (**49**.1)

The eggs are mostly laid from March to August but sometimes in autumn and winter. The shell of the newly hatched larva is about 0.3mm. long and is smooth and transparent. There is a siphonal canal and a characteristic peg-like process on the outer lip of the shell. Larvae of this species are to be found in both inshore and offshore plankton at Plymouth and may be the dominant pelagic gastropod larva of the area (with the possible exception of *Limacina retroversa*). The velum is bilobed at first but soon becomes four-lobed and the border develops a continuous band of brownish pigment. After about eight weeks in the plankton the larva is 0.75mm. long and settles. At this stage the foot is darkly pigmented.

Nassarius incrassatus (**49**.5)

The larvae are commonest in spring and summer and rare in the autumn. Late larvae are abundant in June, July and August, whilst early larvae are found in the winter, spring and summer. They are confined to the inshore plankton.

The newly hatched larva is approximately 0.2mm across the shell and has a colourless bilobed velum which later becomes four-lobed and develops a pigment spot at the tip of each lobe. After a larval life of at least eight weeks the shell is about 0.6mm. long and has a characteristic tooth on the outer lip.

Family TURRIDAE

Mangelia nebula (**50**.1)

The larvae are common in the spring and summer plankton at Plymouth with a maximum from May to September. The shell of the newly hatched larva is approximately 0.3mm. high and the bilobed velum bears a number of orange spots (usually nine to twelve in number). In the later veliger the foot develops a pair of small ciliated lobes and a large median one. The orange spots fuse to form a border or spot at the corner of each velar lobe and the velum becomes enormous and covers the shell and the animal. The shell is tuberculate and striated near the aperture whilst the dark foot is characteristically truncated at its anterior end.

Family APORRHAIDAE

Aporrhais pes-pelicani (**50**.8)

The larva of this species occurs sparingly off Plymouth in May. Only the later larvae are recognizable with certainty, and in these the shell is composed of three whorls of which the last half of the body whorl is striated and the rest smooth. The shell is yellowish and the foot bilobed anteriorly but pointed posteriorly. The most characteristic feature is the enormous six-lobed velum (2.8mm. across). Each lobe bears a brown spot near its tip and the whole velum is bordered with brown. When retracted, the larva resembles that of *Nassarius incrassatus* but the apex to the shell is broader in *Aporrhais pes-pelicani* and the siphonal canal much longer.

Sub-class OPISTHOBRANCHIA

The veligers have sinistrally coiled shells, a feature which distinguishes them from those of the prosobranchs which have dextrally coiled shells.

Order BULLOMORPHA

Philine aperta (**50**.2)

The shell is sinistral and has several whorls. It reaches some 0.3mm in length. The kidney is large and dark in colour and the velum is simple and bilobed.

Order PTEROPODA

Spiratella retroversa (**50**.5)

The most abundant species of thecosomatous pteropod is *Spiratella retroversa.* The eggs are found during most of the year but are commonest in the summer plankton and consist of gelatinous strips some 2mm. long by 0.5mm. broad containing scattered eggs. The larvae hatch after about two days and possess no shell at first but later a symmetrical one about 0.5mm. across develops along with a bilobed velum.

When the shell is about 0.32mm. long it is sinistrally coiled and two rudimentary lappets have formed at the sides of the foot. These rapidly develop into wings and the velum regresses. At about the same time a 'balancer organ' is developed as a projection from the base of the foot. Apart from the brown digestive and anal glands, the larva is colourless. Common in the Channel, and in central and northern N. Sea.

Clione limacina (**49**.6)

This species breeds during the summer and lays planktonic gelatinous egg masses which are about 1mm. long and contain scattered eggs. The newly hatched larva (**49**.6a) is approximately 0.15mm long by 0.1mm broad and the shell is thin and transparent, attached to the larva by longitudinal muscles. The velum is bilobed but is normally retracted in preserved material.

After a period of rapid growth, the larva (**49**.6b) sheds the shell which has acquired striations on the lip, and develops three circlets of cilia. Then the velum is lost and head cones as well as rudimentary wings appear. When the larva reaches about 2mm. in length, the wings are quite large whilst in the 2.8mm. larva all cilia except the last circlet disappear. At 3mm. all the adult organs are formed and the animal is capable of breeding. The head cones and tentacles at this stage are orange-red. the wings pink, and the liver and gonad brownish-yellow.

Order SACOGLOSSA

Limapontia capitata (**50**.3)

The shell has several sinistral coils and reaches 0.35mm in length. The kidney is dark in colour and the simple bilobed velum has dark pigments on the margin.

Order NUDIBRANCHIA

The shells of nudibranchs are sinistral but never have more than 11 coils. There is no pigmentation either in the kidney or velum and eyes are often absent. There is always an operculum and paired statocysts. The larvae of *Aeolidiella glauca* and *Eubranchus pallidus* are shown in **50**.4 and 7, but other species are difficult to distinguish.

IX
Lamellibranch Larvae

THE egg usually hatches as a trochophore which develops into a bivalved veliger, the shell of which (prodissoconch I) has a straight hinge-line. The veliger develops into a veliconcha in which the final larval shell (prodissoconch II) is formed. Finally, after the veliconcha has settled and metamorphosed, the final larval shell is formed. It is not possible to separate the veligers of many lamellibranchs but Rees (1950) has proposed a scheme for the identification of post-veliger larvae whose shape and hinge-structure are known,

The general shape of the larval shell is sufficient to place a larva within a group and for closer identification the position of the ligament and the structure of the hinge is used. Rees has shown that the hinge of lamellibranch larvae is almost diagnostic of the superfamily to which the species belongs and of the eighteen basic types of hinge-structure, fourteen correspond to single superfamilies. The following information is drawn from his paper, and his terminology is adopted here.

The two valves of the shell remain in contact along one side when the valves gape. This region may become thickened to form a *provinculum* which always bears a series of *provincular teeth* whilst hinge-structures at each end of the provinculum form the *lateral hinge system.* In addition, there may be a series of *special teeth* (**51**.14).

Order TAXODONTA

Superfamily ARCACEA

The ligament is nearer the narrow end (measured from the umbo to the ventral surface) and the provinculum bears numerous small teeth equal in size and not clearly separated.

Glycymeris glycymeris (**51**.1)

The larva is very large (360-420μ) and the shell heavy. It possesses the characters of the Arcacea mentioned above. The left valve has a rim around the margin and the right valve a ridge corresponding to the rim. Mainly September-November, but also found from December to January and in summer.

Order ANISOMYARIA

Superfamily MYTILACEA

The provinculum is long and extends beyond the limits of the straight hinge-line region of the shell margin. Teeth are well marked and distinguish the Mytilacea from all other groups (**51**.2a). The two most common genera are *Mytilus* and *Modiolus* which are distinguishable by the texture of the shell as well as by shape.

Mytilus edulis (**51**.2b)

The shell is transparent and about 200-250μ across. The shape is ovoid, an important difference from *Modiolus*. The larvae occur from May to June, but mainly June and July, also November, i.e. mainly in the summer and autumn.

Modiolus modiolus (**51**.5)

The shell is of about the same size as that of *Mytilus edulis* but more massive and with a more pronounced umbo. The ribs and coloration are also more obvious. The shape is also different from that of *Mytilus edulis*; in *Modiolus modiolus* the broad end (measured from the tip of the umbo to the ventral surface) droops ventrally. Found in the plankton from June to December, but mainly in October and November, i.e. mainly in autumn and winter (cf. *Mytilus*).

Superfamily PECTINACEA

This contains the genera *Chlamys* and *Pecten* as well as *Lima*. The provinculum is similar to that of the Mytilacea but the teeth are extremely small in the middle region (**51**.3).

Lima sp. (**51**.4)

The larva is large, some 310μ long, and of a quite characteristic triangular shape.

Superfamily ANOMIACEA

The provinculum is the same as in the Pectinacea but the shell is of a characteristic shape, e.g. *Anomia ephippium* (**51**.7). The shell is approximately 250μ long.

Order EULAMELLIBRANCHIA

Sub-order HETERODONTA

Superfamily CARDIACEA

The provinculum of the left valve is thicker than that of the right and has a number of gaps along its edge into which project the spiky teeth of the right provinculum (**51**.8). Some species of *Cardium* have a high umbone whilst others have a low umbone, but all are of a similar general appearance and all possess an obvious pallial groove some distance from the edge of the shell.

Cardium ovale (**51**.9)

The length of the shell is variable: it is normally about 240μ but may reach 340μ. The umbones of the shell are high but the valves are not so convex as in other species.

Cardium edule (**51**.12)

The shell is about 160μ in the three-week larva and reaches 300μ in the five-week larva. The shell is round and smooth, ribs appearing only after metamorphosis. The 1mm. stages are common in July.

Cardium echinatum (**51**.10)

The larva of this species is much larger at metamorphosis than *C. ovale*, *C. edule* or *C. scabrum*, and reaches 480μ in length. The valves are very convex, giving a globular appearance to the larva. Found in spring, summer and autumn plankton.

Laevicardium crassum (**51**.13)

The larva of this species is also large, but the valves are not so convex and are truncated posteriorly. The smallest veliger is 200μ, reaching 450 in length at metamorphosis. Common in the spring and summer plankton.

Cardium scabrum (**51**.11)

The shell of the larva of this species bears characteristic concentric striations and reaches 160μ in length before metamorphosis. The shell is more oblong than *C. edule*.

Superfamily VENERACEA

The shells are always ovoid and rather small. The provinculum has distinct lateral ridges at each end forming part of the lateral hinge system. These ridges extend almost halfway ventrally at the posterior end and about one-third of the way at the anterior end (**52**.1).

Venus striatula (**52**.3)

The shell bears a number of distinct concentric grooves. The dorso-lateral margin of the narrow end is convex - a feature which distinguishes it from *Venus ovata*. The length is about 220μ. The larvae occur between October and November, being most common in November.

Venus ovata (**52**.2)

The larva is essentially similar in structure to that of *V. striatula* but the dorso-lateral margin of the narrow end is straight. The length is normally about 220μ and the larvae occur in September and October but mainly in October, i.e. autumn and winter.

Dosinia sp. (**52**.4)

The shell is thicker dorso-ventrally, the broad end not curving away so steeply from the umbo as in other venerids. This results in the dorsal edge being longer. The length of the shell is normally about 220μ.

Petricola pholadiformis (**52**.5)

The larva has a thicker shell than those of other venerids and possesses a strong provincular projection. The valves are more strongly coloured than in other venerids. The length is usually about 170μ.

Superfamily MACTRACEA

The provinculum of all species belonging to this superfamily, except *Mactra corallina*, is characterized by a series of rectangular teeth of which one on the posterior end of the right valve is especially prominent, being enhanced by the suppression of the other teeth in some species. The prominence of this tooth is a diagnostic and readily recognizable feature of the Mactracea (**52**.9).

Mactra corallina (**52**.6)

The broad end of the shell is truncated and the whole shell is of a yellowish colour. The umbones are characteristically indistinct and the valves are crossed by fine concentric striations and occasional coarser lines. The shell is about 200μ long and the larvae are found in the S.W. Dogger area from January to March and also in October but chiefly in April and again in September, i.e. a major occurrence in spring and a minor one in autumn. The provinculum is not toothed as in other members of the superfamily.

Spisula solida (**52**.7)

The shell is brownish and crossed by coarse concentric lines. It is quite unmistakable and may reach 360μ. Occurs in May and April and again in September.

Spisula elliptica (**52**.8)

The shell of this species has a truncated broad end as in *Mactra corallina* but the high umbones distinguish it from this species, as also does the presence of the peg-like process on the provinculum. The larva reaches about 350μ in length and the yellowish shell is crossed by groups of fine concentric lines. There is a minor outburst of larvae in May and a major one in August. Also found in the plankton during September, October and November.

Superfamily TELLINACEA

The characteristic feature of the provinculum of larvae of this superfamily is that the ligament is in the *centre*, rather than at one end, of the provinculum (**52**.11). There are a number of small teeth resembling the taxodont condition. The ridges and flanges alternate on each valve whilst special teeth arise in later larvae.

Tellina fabula (**52**.12)

The antero-dorsal margin of the shell is concave and the shell is crossed by fine concentric lines. The larvae are of infrequent occurrence in the plankton during July, August and September.

T. crassa (**52**.10)

The shell is crossed by a number of deep concentric lines which become more widely spaced towards the edge. These coarse striations are diagnostic of *T. crassa*. This species occurs only during September and only in the southern N. Sea.

Sub-order ADAPEDONTA

Superfamily SOLENACEA

Larvae of species belonging to this group are characterized by the possession of an *external* ligament which appears in the 280μ larva and becomes larger with the increasing size of the larva. In addition, the right umbo is higher than the left.

Cultellus pellucidus (**52**.13)

The postero-dorsal edge of the shell is concave whilst the left umbo is shorter than the right. The length is approximately 330μ.

Ensis ensis (**52**.14)

The broad (posterior) end is sharply truncated and the umbones are indistinct, the left being lower. The larvae occur mainly in April and May but also in June and July. The length of the shell is approximately 250μ.

Ensis siliqua (**52**.15)

The posterior end is not so truncated as in *E. ensis* and the umbones are more pronounced. The larvae occur at the same time as, but are rarer than, those of *E. ensis*. The length of the shell is approximately 230μ.

X
Ectoproct Larvae

AS IN many animals, the length of planktonic larval life in the Ectoprocta is variable and in most is omitted. Those which are lecithotrophic and which lack a gut may spend only some twenty-four hours afloat and are of little importance in the plankton but may be temporarily abundant. The planktotrophic type of larva, or cyphonautes, is a ciliated larva and has some features in common with a trochophore. It is flattened from side to side and triangular in shape, and there is a bivalved calcareous shell. The main band of swimming cilia is ventrally situated and there is a dorsal apical organ. A ∩-shaped ciliated gut is present.

Several kinds of ectoproct larva occur in our plankton but not all have been related to the adult species (See Ryland 1965). Only two kinds of fully-formed cyphonautes are likely to be common. These belong to *Electra pilosa* and *Membranipora membranacea.*

Electra pilosa (**54**.9)

The larva is of a conical shape with a blunt apex which is displaced somewhat posteriorly. The colour is an opaque yellow-brown, and the width is approximately 0.5mm. at the widest point of the later larva. The larvae of this species, sometimes called *Cyphonautes compressus*, constitute almost half of the cyphonautes to be found in the plankton and are especially common in the winter from October to January.

Membranipora membranacea (**54**.10)

The cyphonautes of this species is distinguished from that of *E. pilosa* by its transparency and more nearly symmetrical shape. The posterior margin of the oral edge bears a rounded protuberance. The length of the oral edge is 0.7mm., i.e. much larger than that of *E. pilosa*. The cyphonautes of *M. membranacea* is the dominant cyphonautes larva of the summer and reaches its maximum abundance during October and November. It is sometimes referred to as *Cyphonautes balticus*.

XI
Echinoderm Larvae

ALMOST all echinoderms have a pelagic larval stage in their development which, as in other animals, tends to be long or short according to whether the egg is poorly or well supplied with yolk.

In all echinoderms the first larval stage is a ciliated gastrula which soon transforms into a bilateral larva termed a dipleurula. From this stage there is a great diversification, each class of echinoderms having its own characteristic type. Nevertheless all these types have been shown to be referable to the dipleurula pattern and derived from it by relatively simple modifications. The larvae so formed are given special names. The asteroids have a *bipinnaria* which later develops into a *brachiolaria* before the adult form is attained. The ophiuroids and echinoids both have a larva called a *pluteus*, that of the ophiuroids being called an *ophiopluteus* and that of the echinoids an *echinopluteus*. The larva of the holothuroids is an *auricularia* which later metamorphoses into a *doliolaria*, and finally into a miniature holothurian called a *pentacula* larva. All these larval types are at times abundant in the plankton particularly in the spring, of which season they are good indicators, so that it is of interest to distinguish between the more common species.

In addition to these larvae, post-larval echinoderms often leave the bottom and are sometimes caught in summer plankton hauls; these will not be considered here.

Class CRINOIDEA

The only British crinoid is *Antedon bifida* which has an ovoid elongated larva, *larva en tonnelet* (barrel-like), which is surrounded by four hoops of cilia (**53**.4), and bears in the mid-line on the ventral surface an adhesive pit anteriorly and a stomodaeal invagination posteriorly. An apical plate, bearing a tuft of cilia and underlain by nervous tissue, is present. The skeleton is well developed and consists of a number of perforated calcareous plates; in this respect it differs from the *larva en tonnelet* of holothurians in which the skeleton is in the form of small plates around the anus rather than distributed throughout the larva as in *Antedon*.

Class ASTEROIDEA

There are two larval forms in the Asteroidea, the bipinnaria and the brachiolaria.

BIPINNARIA

This has no skeleton and may be regarded as having arisen from the dipleurula by the further growth of the two antero-lateral pre-oral lobes which have then fused in the median plane at their distal ends. The band of cilia, which in the auricularia is a continuous ring, later becomes split into two independent rings. This enables the two larval forms to be distinguished.

The sides of the larva become drawn out into symmetrical lobes called arms which are often very long and bear the lateral extensions of the ciliated field. These are twelve in number and consist of unpaired ventro-median and dorso-median arms, paired postero-orals, postero-dorsals, postero-laterals, antero-dorsals and pre-orals. Bipinnariae of different genera are distinguishable by their general proportions.

Astropecten (**53**.3)

The bipinnariae are perhaps the least modified from the dipleurula condition, the lateral arms being short. But a characteristic feature - the circum-oral ciliary band split into two - can be seen.

Luidia sarsi (**53**.1)

The pre-oral lobe is elongated and the medio-dorsal arm is also long. These differences from *Astropecten* result in the larva having a quite distinctive shape. The larvae are in the plankton from August to October, and reach 2.5-3.0cm. in length.

Asterias rubens (**53**.2)

All the arms in this species are extremely elongated from an early stage so that the bipinnaria is easily distinguished from those mentioned above. The larvae are about 2mm. long and occur in the plankton from May to September.

BRACHIOLARIA

The brachiolaria stage is one in which the organs for fixation are developed. These organs are necessary for metamorphosis; they are composed of three special appendages in the pre-oral region called brachiolarian arms. These are armed with numerous adhesive papillae but except for the medio-dorsal one, which sometimes bears a portion of the circum-oral ciliary band, they are unciliated.

Class OPHIUROIDEA

The characteristic larva of the ophiuroids is the ophiopluteus which is a conical larva, the apex of the cone corresponding to the posterior end of the larva, bearing four pairs of arms. These are termed the antero-lateral, postero-lateral, postero-dorsal and postero-oral arms; that is, there are neither pre-oral arms nor antero-dorsals. The postero-laterals are very much longer than the other arms and give the larva a quite unmistakable appearance, being always long and upwardly pointed.

There are a number of different ophioplutei which can be separated chiefly on a basis of the relative length of the postero-lateral arms.

Ophiothrix fragilis (**53**.10)

The postero-laterals are enormously long and the body is about 0.3mm. long. Ophioplutei of this species are often abundant in the plankton from April to August.

Ophiocomina nigra (**53**.9)

The postero-laterals are not long and bear a pair of vibratile appendages at their base.

Amphiura filiformis (**53**.6)

The ophiopluteus of this species is characterized by the complete absence of postero-dorsal arms. Common in the early summer plankton

Ophiura albida

This has large arms whilst in the ophiopluteus of *O. textura* the arms are very short.

Metamorphosis occurs whilst the ophiopluteus is still pelagic. The arms shorten whilst their skeletal supports degenerate. The anus and intestine degenerate whilst the adult skeleton is formed and the larval skeletal rods are absorbed. Finally, the increased weight of the skeleton, coupled with the shortening of the arms, causes the larva to sink and adult life is taken up.

Class ECHINOIDEA

The echinopluteus is the only larva of the echinoids and resembles the ophiopluteus except that the body is compressed laterally rather than dorso-ventrally. Also, in many echinoplutei the apex of the ventral cone is prolonged into an apical appendage.

There is the same arrangement of arms as in an ophiopluteus: antero-lateral, postero-lateral, postero-dorsal and postero-oral ones, but the dorsal half bears an additional pair, the antero-dorsal arms, which are missing entirely in the ophio pluteus but are present in the bipinnaria of asteroids. In addition, the echinopluteus may bear a pair of characteristic pre-oral arms, e.g. in *Echinocyamus pusillus* (**53**.7). This species is 0.5mm long and occurs from the middle of March to September.

Whereas in ophioplutei the postero-laterals are always long and directed upwards, those of the echinopluteus may be absent or, if present, point laterally or even posteriorly as in the pluteus of *Echinocardium cordatum* (**53**.8). The pluteus possesses a long apical appendage. The arms are swollen at their proximal ends, and the postero-lateral arms point to the rear. The larvae of this species occur in the plankton from July to September and are about 0.6mm. long. The postero-orals, postero-dorsals, antero-laterals and antero-dorsals are long.

Psammechinus miliaris (**54**.1)

The echinopluteus is 0.5 to 1mm. long and is characterized by the absence of postero-laterals. It occurs in the summer plankton.

Echinus esculentus

The echinopluteus bears ciliated structures called epaulettes at the base of the postero-dorsal arms which are orientated transversely to the vertical axis of the larva. It is about 1.5mm. long and occurs in the spring and summer plankton.

Spatangus purpureus (**54**.2)

The postero-oral arms are long and, like the echinopluteus of *Echinocardium cordatum*, there is a long apical appendage. The length from the tip of the apical appendage to the tip of the postero-oral arms may reach 6mm. The larvae are common in the late spring and early summer plankton. Postero-lateral arms are absent.

Class HOLOTHUROIDEA

The holothurians have a number of larval stages before the adult form is attained. In the family Holothuriidae there is a typical auricularia larva (**53**.3), but in the Dendrochirota and also in *Leptosynapta inhaerens* the auricularia stage is omitted, and a *larva en tonnelet* (barrel-like) is formed (**54**.4, 6). This resembles the doliolaria stage which follows the auricularia in a typical developmental sequence. Both the doliolaria and the *larva en tonnelet* develop into the final larval stage, the pentacula (**54**.5), which finally settles on the substratum and reaches the adult stage.

AURICULARIA

The auricularia larva is the least modified larval form and resembles the basic dipleurula type. The circum-oral ciliary band of the dipleurula has become drawn out into paired pre-oral lobes anteriorly and paired anal lobes posteriorly. Clearly, as in the dipleurula, the circum-oral ciliary field is continuous.

LARVA EN TONNELET

This larva is characteristic of the dendrochirotes (e.g. *Cucumaria*) (**54**.4). The ovoid body has none of the features which characterize larvae derived from the dipleurula since both ventral depression and circum-oral ciliated band are absent. Instead, there is a series of transverse ciliary rings which give the larva a similar appearance to the doliolaria larva of those holothurians which pass through an auricularia stage. The number of rings varies with the species (e.g. four in *Leptosynapta* (**54**.6) and five in *Cucumaria*). There is a pre-oral lobe in some species, whilst there is a stomodeal vestibule ventrally and a terminal anus. Finally, the primary body cavity is considerably reduced and is filled with complex viscera. This larva is rarely found in the plankton since it swims near the sea-floor.

DOLIOLARIA

Unlike the pluteus and bipinnaria, the auricularia is a primary larva and metamorphoses into a secondary larva of a quite different appearance. This is the doliolaria, so called because of its superficial resemblance to the pelagic tunicate *Doliolum*. This larval stage is extremely similar to the *larva en tonnelet* of the dendrochirotes and is derivable from the auricularia by certain important modifications. The ventral depression and lateral ciliated lobes of the auricularia are lost whilst a terminal prebuccal vestibule is developed.

The ciliary band of the auricularia becomes fragmented into 16 isolated pieces which are reorientated into six transverse parallel bands. The first ring becomes incorporated into the vestibule so that five rings only are visible externally; at the same time, two small bands of nervous tissue fuse to form a ring. The doliolaria can be distinguished from the *larva en tonnelet* by the absence of a pre-oral lobe and by the terminal position of the prebuccal vestibule.

The doliolaria and the *larva en tonnelet* both develop into the final larval stage, the pentacula. But in reality the beginning of metamorphosis is foreshadowed in the doliolaria and *larva en tonnelet*, both of which have many features in common with the final larval stage and with the adult.

PENTACULA

The pre-oral lobe disappears completely whilst the five peribuccal appendages hidden inside the vestibular cavity emerge and the larva comes to look like a small holothurian. A pair of podia is developed posteriorly on the ventral surface. The larva continues its planktonic existence but soon ciliated bands disappear and the larva attaches itself by its podia (**54**.5).

XII
Enteropneust Larvae

Phylum HEMICHORDATA

HYMAN (1959) and most recent authors agree that the hemichordates are sufficiently distinctive a group to warrant the rank of a separate phylum, although earlier workers placed them in the phylum Chordata. On certain shores around our coasts Enteropneusta are exceedingly common but most belong to the genus *Saccoglossus* whose lecithotrophic larva rarely, if ever, appears in the plankton. A few British species, however, for example *Balanoglossus clavigerus*, have a planktotrophic larva with a long pelagic life termed a tornaria (**54**.7). Other tornariae are also taken in some sea areas but many cannot with certainty be related to the adult worms. Moreover, the structure of the tornariae varies a good deal with age and the younger ones can easily be confused with bipinnaria and auriculariae. Once the apical tuft and telotroch (ciliary girdle, girdles) round the hinder end of the body have developed - features not found in echinoderm larvae - confusion is not likely. An excellent summary of the features and distribution of various tornariae is given by Burdon-Jones (1957).

XIII
Fish Eggs and Larvae

THE great majority of teleost fishes shed their eggs and sperms into the surrounding water and the resultant fertilized eggs may float freely in the plankton. In other species, however, such as the herring, the blennies, sucker fish and many other inshore species, the eggs are stuck to objects on the sea floor. Some species even mount guard over the eggs or incubate them in a brood pouch as in pipe fishes. The large-yolked eggs of elasmobranchs are laid in horny capsules which bear coiled filaments used for attaching the eggs to algae, to stones or for anchoring them in deposits. These, as well as those teleost eggs attached to the sea bed, are called 'demersal eggs' and will not receive further mention since they do not contribute to the plankton. It does not follow, however, that larvae hatching from such demersal eggs remain on the sea bed. Most of them become plankontic and only at a later stage may seek the bottom; planktonic fish larvae may thus be derived from either planktonic or demersal eggs.

The identification of planktonic fish eggs is not easy, especially in preserved material, for this often leads to a loss of pigmentation from the embryo and yolk sac. In general, the presence or absence of oil globules, whether the yolk is segmented or homogeneous, the size of the perivitelline space, the size and shape of the egg and the nature of its surface membrances all aid in the identification of fish eggs. In fresh material, however, pigmentation of the embryo, yolk sac and eyes all form important criteria for identification. Some of the features of the common pelagic fish eggs are summarized in the table on pages 134-136, but reference should be made to Russell (1976) before any firm identification is made.

The newly hatched larva is usually less than 4mm long and during this phase of development the yolk sac is used as a source of food. By the time the yolk sac has finally disappeared, the mouth and anus have become fully functional and the young fish then enters the post-larval phase during which adult characters appear. The young larvae are often so unlike the adults that specific identification is extremely difficult, but by the end of the larval period the pigment cells occupy positions characteristic of the post-larva and make identification easier. The earliest post-larval pigmentation is usually characteristic of the species and this persists until the fish becomes silvery and fin rays, vertebrae and ribs ossify to give the structural features by which the adult is identified. Eggs and larvae of some of the commoner fishes found in North European waters are illustrated in **55** and **56.**

Larvae and Post-Larvae

Family CLUPEIDAE (**55**. 1-6)

Clupeid larvae are characteristically elongated with the anus near the posterior end of the tail. The newly hatched larva of the herring (*Clupea harengus*) is 6-9mm long and is much larger than the larvae of the sprat, pilchard and anchovy which are only 3-4mm long. The pilchard (*Sardinia pilchardus*) can be distinguished by the presence of an oil globule, whereas the two without oil globules are distinguished by the fact that in the anchovy (*Engraulis encrasicolus*) the yolk is elongated whereas in the sprat (*Sprattus sprattus*) it is spherical.

Family BELONIDAE

Belone belone, garfish (**55**. 7-8)

The larva is characterized by having the lower jaw projecting slightly beyond the upper jaw and by being covered with numerous yellow and black chromatophores. The adult garfish has elongated upper and lower jaws and during the post-larval phase the lower jaw is considerably longer than the upper. This is known as the 'rhamphistoma stage' and by the time the young fish has reached 90mm the lower jaw may be as much as 18mm in length. Larval and post-larval stages occur in the plankton in June and July.

Family MERLUCCIIDAE

Merluccius merluccius, hake (**55**. 9-11)

The newly hatched larva is approximately 3.0mm long and is immediately distinguishable by the three post-anal stellate melanophores, and by the large pigmented oil globule at the posterior end of the yolk sac. The post-larva is characteristically deep dorso-ventrally and has conspicuous pigmentation on the head and pectoral region as well as the three post-anal bars. Post-larvae occur in European waters from June to November.

Family GADIDAE

Gadus morhua, cod (**55**. 12-13)

The newly hatched larva is approximately 4.0mm long and has pigmented eyes, two post-anal bars of melanophores as well as several ventral caudal melanophores. There is no oil globule. The lower jaw of the post-larva becomes distinctly angular and at 9mm the first rays appear in the dorsal and anal fins. Post-larvae occur in spring and early summer, but once a length of 20-30mm is reached, they become demersal.

Melanogrammus aeglefinus, haddock (**55**. 14)

The newly hatched larva is 3-4mm long and is characterized by an absence of dorsal pigmentation in the caudal region, a feature which persists until the post-larva reaches 15-16mm and which distinguishes the larva from other gadoid species. There is a single row of melanophores ventrally as well as scattered pigment cells on each side in the pectoral region and on the head. Soon after yolk absorption, yellow pigment appears anteriorly and the pectoral melanophores form a row which extends back to the posterior margin of the pectoral fins. Larvae and post-larvae occur in the plankton in May and June in north European waters.

Trisopterus luscus, bib or pout (**55**. 15)

The newly hatched larva is some 3mm long. Both larvae and post-larvae are characterized by the absence of black pigmentation in the last third of the post-anal region. The body is yellowish in colour due to the presence of small yellow chromatophores over much of the body. Post-larvae occur throughout the year in European waters but are commonest from October to May.

Trisopterus minutus, poor cod (**55**. 16)

The larva is only 2.5mm long and covered with small yellow chromatophores. Pigmentation of the post-larva is distinctive and consists of dorsal and ventral rows of about nine large melanophores which extend as far as the caudal fin. The head is largely free of melanophores except for one on each side above the eyes. Post-larvae are common in the plankton from March to May.

Merlangius merlangus, whiting (**55**. 17)

The newly-hatched larva is 32-3.5mm long and has yellow pigment on the yolk sac and primordial fin. The yolk is absorbed when the larva reaches 4mm and at this stage the fin has only reddish-yellow pigment. Dorsal and ventral rows of melanophores are present, but the dorsal row does not extend as far posteriorly as the ventral row. This feature distinguishes the early larva from that of *T. minutus* (poor cod). Later development of pigmentation involves an increase in melanophores on the head and sides of the body. By the time the post-larva reaches 11mm a few melanophores are to be found between the rays at the base of the anal fins. This feature serves to distinguish young whiting from other post-larval gadoids. Post-larvae occur from March to July.

Pollachius virens, coalfish

The larva is 3.4-3.8mm long and has two post-anal bars of pigment, but the caudal region has no melanophores. This last feature distinguishes the larva from that of the cod *G.morhua*. Post-larvae occur in April and May.

Molva molva, ling (**55**. 18-19)

The larva is characterized by the heavily pigmented oil globule and the greenish-yellow pigment on the body, primordial fin and yolk sac. Post-larval pigmentation comprises two post-anal pigment bars as well as a row in the pectoral region and above the head. The post-larva is characterized by the long pelvic fins, each with three rays between which the membrane is pigmented. Larvae and post-larvae occur in the plankton between April and July.

Gaidropsarus sp and *Ciliata* sp, rocklings

The larvae of rocklings are very small, usually hatching at 1.8-2mm in length. They all have a pigmented oil globule and two or more post-anal pigment bars in the caudal region. The post-larvae all have large pigmented pelvic fins. At 10-12mm length the body becomes silvered and shoals of these 'mackerel midges' may be seen in shallow waters. Post-larvae may be found throughout the year but are most common in April and May off Plymouth.

Family CARANGIDAE

Trachurus trachurus, horse mackerel (**55**. 20)

The newly hatched larva is 2.5mm long and is characterized by the position of the pigmented oil globule which lies in the anterior part of the segmented yolk. At first there are melanophores and brown-yellow chromatophores irregularly distributed over the body but later the pigment is reorganized into dorsal and ventral rows of melanophores with brown pigment on the edges of the dorsal and ventral fins. A swimbladder appears when the larva is only 4mm long. Post-larvae occur in the plankton during the late summer from July to October.

Family MULLIDAE

Mullus surmuletus, red mullet (**55**. 21)

The larva has an ovoid yolk sac which projects well beyond the head and has an anterior oil globule. Black peritoneal pigment is heavy in the post-larva, which is distinctly silvery by the time the post-larva is 8mm long. Post-larvae occur in the summer from June to August.

Family AMMODYTIDAE

Ammodytes, Gymnammodytes, Hyperoplus sp, sandeels (**55**. 22)

The larvae and post-larvae of sandeels superficially resemble clupeids, but in the latter the anus opens near to the caudal end whereas in sandeels it opens near the middle of the larva. Pigment tends to be confined to dorsal and ventral melanophores on the body. There are also rows of large melanophores along the dorsal and ventral parts of the alimentary canal. Larvae and post-larvae occur in the plankton from May to November.

Family TRACHINIDAE

Trachinus vipera, weaver (**56**. 1-2)

The pelvic fins are prominent and coloured black and yellow in the newly hatched larva. Apart from in the earliest larva, the melanophores aggregate on the body to form a wide anal bar and a narrow one midway along the post-anal part of the body. These characteristic bands disappear in the post-larva and are replaced by heavy pigmentation in the peritoneum and on the well-developed pelvic fins. Post-larvae occur from May to September.

Family SCOMBRIDAE

Scomber scombrus, mackerel (**56**. 3-4)

The newly hatched larva is 3.3-3.9mm long and has melanophores on the head as well as double rows of melanophores dorsally and ventrally in the posterior part of the body starting some way behind the anus. Yellow pigment occurs behind the eyes and on the oil globule. There is thus a pigment-free zone near the anus and this persists in the post-larva. Post-larvae occur from April to July.

Family CALLIONYMIDAE

Callionymus lyra, dragonet (**56**. 5-6)

The larva is very small and is only 2.3mm by the time the yolk is absorbed. The large high head and short body are characteristic. The post-larva is heavily pigmented and by the time it reaches about 4mm in length the diagnostic preopercular spine has appeared. Post-larvae are extremely common in the plankton from April to August and may dominate the catch during May and June off Plymouth.

Family BLENNIIDAE

Blennius sp, blenny (**56**. 7-9)

Larvae and post-larvae of blennies are characterized by strongly pigmented pectoral fins. In *B. pholis* the fin is elongated whereas in *B. ocellaris* it is rounded. In *B. gattorugine* the pectoral fin is well-developed and pigmentation is especially heavy in the peritoneum. Post-larvae occur in the plankton from May to September.

Family TRIGLIDAE

Eutrigla gurnardus, gurnard (**56**. 10)

The newly hatched larva is 3-4mm and is covered with black and yellow chromatophores which also occur in the yolk sac and oil globule. The shape of the head and large pectoral fins with radial marginal melanophores is diagnostic of post-larvae of the gurnard. They occur in the plankton during the summer months from May onwards.

Family BOTHIDAE

Scophthalmus maximus, turbot (**56**. 11-13)

The newly hatched larva is 2-2.8mm long and covered with stellate red and black chromatophores. An obvious post-anal bar occurs where the chromatophores extend onto the primordial fin. The post-larva is rather deep dorsoventrally and is reddish in colour except for the caudal region. Larvae and post-larvae of the turbot occur in the plankton from June to September.

Scophthalmus rhombus, brill

The newly hatched larva is larger than that of the turbot and is about 4mm long. The body is uniformly covered with orange-yellow and black chromatophores and there is a post-anal bar as in the turbot. The post-larva is distinguishable by the orange-yellow colouration as opposed to the red colour of the turbot. Larvae and post-larvae of the brill occur in April to September.

Family PLEURONECTIDAE

Limanda limanda, dab (**56**. 14)

The newly hatched larva is about 2.7mm long and is covered mainly with yellow chromatophores anteriorly and black ones posteriorly. At a later stage the yellow pigment predominates dorsally and the black ventrally with black and yellow chromatophores appearing on the pectoral fin. At this stage the primordial fin becomes pigmented and in the post-larva there are four or five groups of melanophores on the dorsal fin and three on the ventral fin. Post-larvae occur in April and May.

Pleuronectes platessa, plaice (**56**. 15-16)

The larva is 6-7.5mm long and is much larger than that of the dab or flounder. The dorsal part of the body and yolk sac are covered with canary-yellow chromatophores whilst melanophores predominate in the body ventrally. There are no melanophores on the pectoral fin in contrast to *Limanda*, and metamorphosis begins when the post-larva reaches 12mm. Post-larvae occur from January to May.

Platichthys flesus, flounder

The larva is 2.3-3.3mm and has chrome-yellow pigment. It rather resembles that of *Limanda* but is more closely and brightly pigmented. Post-larvae occur in the plankton from March to June.

Family SOLEIDAE

Solea solea, common sole (**56**. 17-18)

The newly hatched larva is 2.5-3.75mm long and is heavily pigmented with stellate chromatophores over the yolk, body and primordial fins which project somewhat over the head. Larvae and post-larvae occur in March and June but are never very common.

Family LOPHIIDAE

Lophius piscatorius, angler fish (**56**. 19)

The newly hatched larva is large and is 4.5mm long with the rudiments of the pelvic fins arising near the upper surface of the yolk sac. Long filamentous dorsal fin rays develop early and much of the head region and yolk sac is heavily pigmented. Post-larvae are rather rare but may be caught from March to September.

TABLE SHOWING THE FEATURES OF COMMON PELAGIC FISH EGGS

Species	*Diameter of egg (mm)*	*Diameter of oil globule (mm)*	*Spawning period N. European waters*	*Features*
Family CLUPEIDAE				Thin egg capsules. *Yolk segmented*
Sardinia pilchardus (pilchard)	1.3-1.9	0.14-0.18	June-July, Oct.-Nov.	Large perivitelline space and oil globule
Sprattus sprattus (sprat)	0.8-1.3	none	Feb.-March to early summer	Thin egg capsule. No oil globule
Engraulis encrasicolus (anchovy)	1.2-1.9 x 0.5-1.2	none	rare	Ovoid egg. No oil globule
Family BELONIDAE				
Belone belone (garfish)	3-3.5	none	May-July	Unsegmented yolk. Egg membrane with filamentous processes
Family MERLUCCIIDAE				
Merluccius merluccius (hake)	0.94-1.03	0.25-0.28	June-Aug.	Embryo, yolk sac and large oil globule black pigmented

TABLE SHOWING THE FEATURES OF COMMON PELAGIC FISH EGGS (CONT.)

Species	*Diameter of egg (mm)*	*Diameter of oil globule (mm)*	*Spawning period N. European waters*	*Features*
Family GADIDAE				Thin egg capsules. Yolk unsegmented. Usually no oil globule
Gadus morhua (cod)	1.16-1-89	none	Feb.-May	Embryo black pigmented. Late stage has two post-anal pigment bars and melanophores in tail
Melanogrammus aeglefinus (haddock)	1.2-1.7	none	Jan.-May	Post-anal ventral row of melanophores and unpigmented dorsal surface except above pectoral fin
Trisopterus luscus (bib or pout)	0.9-1.23	none	Oct.-May	Melanophores over much of body but eyes hardly pigmented
Trisopterus minutus (poor cod)	0.95-1.03	none	Feb.-June	Head and body of embryo and yolk sac with *yellow* as well as black pigment
Merlangius merlangus (whiting)	0.97-1.32	none	Feb.-June	Early stages with black pigment only. Later stages with *yellow* on yolk sac and fins. Like poor cod
Molva molva (ling)	0.97-1.13	0.28-0.31	April-June	*Oil globule black pigmented.* Fins, yolk sac and ventral body yellow/green. Melanophores in two rows down body
Ciliata mustela (five-bearded rockling)	0.66-0.98	0.11-0.16	March-May	Small eggs. Melanophores on oil globule but not on yolk. Two patches of post-anal chromatophores
Gaidropsarus sp (three-bearded rockling)	0.78-0.84	0.14-0.16	Jan.-Aug.	Late embryo with only one post-anal melanophore patch
Family CARANGIDAE				
Trachurus trachurus (horse mackerel)	0.8-1.0	0.19-0.28	May-Sept.	Segmented yolk. Oil globule and contours of body with brown/yellow and black pigment
Family MULLIDAE				
Mullus surmuletus (red mullet)	0.81-0.91	0.23-0.25	May-July	Oil globule very large. Yolk segmented at edges. Melanophores onyolk, posterior part of oil globule and on lateral body
Family TRACHINIDAE				
Trachinus vipera (weaver)	1.0-1.37	many small droplets	May-Aug.	Unsegmented yolk with 6-30 greenish oil globules. Large yellow cells initially with more black cells later

TABLE SHOWING THE FEATURES OF COMMON PELAGIC FISH EGGS (CONT.)

Species	*Diameter of egg (mm)*	*Diameter of oil globule (mm)*	*Spawning period N. European waters*	*Features*
Family SCOMBRIDAE				
Scomber scombrus (mackerel)	1.0-1.38	0.28-0.35	March-Aug.	Unsegmented yolk. Yellow pigment behind eyes and near attachment to yolk sac. Black pigment cells on front half of oil globule
Family CALLIONYMIDAE				
Callionymus lyra (dragonet)	0.81-0.97	none	Jan.-Aug.	Yolk with peripheral segmentation. Surface of egg with hexagonal marks
Family TRIGLIDAE				
Eutrigla gurnardus (gurnard)	1.27-1.55	0.25-0.33	Jan.-Aug.	Unsegmented yolk with large oil globule. Black and yellow pigment on fins, yolk sac and oil globule. Pectoral fins large
Family BOTHIDAE				
Scophthalmus maximus (turbot)	0.91-1.2	0.15-0.22	April-Sept.	Unsegmented yolk. Egg membrane striated. Embryo and oil globule with *red/brown* and black cells
Scophrhalmus rhombus (brill)	1.24-1.5	0.16-0.25	April-Sept.	Unsegmented yolk. Embryo and yolk with *yellow* cells. Egg yellowish and larger than turbot. Late stage becomes orange yellow
Family PLEURONECTIDAE				Pelagic eggs with no oil globules. Unsegmented yolk
Limanda limanda (dab)	0.66-1.2	none	Jan.-Sept.	Egg of glass-like transparency. Late embryo with *lemon yellow* pigment anteriorly
Pleuronectes platessa (plaice)	1.66-2.17	none	Dec.-March	Finely striated and iridescent egg membrane. Large egg with vivid yellow embryo
Platichthys flesus (flounder)	0.8-1.13	none	Jan.-July	*Chrome yellow* pigment
Family SOLEIDAE				
Solea solea (common sole)	1.0-1.6	many small aggregated droplets	Feb.-June	Yolk with peripheral segmentation. Clusters of tiny oil droplets on ventral side of embryo. Embryo and yolk with yellow and Black chromatophores
Family LOPHIIDAE				
Lophius piscatorius (Angler fish)	2.3-3.1	many oil globules 0.5-0.9	Feb.-Aug.	Large oval eggs in gelatinous ribbons 10m long x 30cm wide. Yolk yellow

XIV
A Brief Account of the Hydrography of the British Isles and Adjacent Seas

THE waters around our coasts and adjacent sea areas are derived from three main sources. First, water of high salinity and temperature, but not very dense, brought across to Europe by the N. Atlantic Current or Drift, itself an important part of the Gulf Stream system; second, fresh water from the Baltic and great rivers; third, cold bottom water from the Norwegian Sea. Of these components, the first is, perhaps, the most important. It is far from being a simple streaming of sub-tropical water reissuing from the Gulf of Mexico, although there can be little doubt that the Florida current streaming out of the Gulf through the Strait of Florida northwards to Cape Hatteras, and the Gulf Stream proper, from Cape Hatteras to the Grand Banks, are directly responsible for a system of east-going currents which traverse the Atlantic and which are now usually called the N. Atlantic Current. This has several branches and the course of all of them is not known. Most of the water flows north-easterly but an important tongue, the Irminger Current, turns back westward south of Iceland, leaving the rest of the northerly stream as the Norwegian Current to pass over the Wyville-Thompson Ridge into the Norwegian Sea, finally to reach the Polar Sea itself. More southerly branches of the N. Atlantic Current seem mostly to end in great eddies near the European coasts. It is a south-going stream from the main Norwegian Current which, passing mainly to the east of the Shetlands, enters the N. Sea from the north, supplying it with warm Atlantic water. There it slowly mixes with less saline waters from the deeper parts of the Norwegian Sea. The N. Sea also receives a lesser amount of Atlantic water, partly mixed with coastal and Channel water which has passed eastwards through the Channel and Strait of Dover, and enters the Southern Bight. In some years this inflow of Channel water is insignificant and its flow is apparently related to the amount of Atlantic water entering the N. Sea from the north. Moreover, the current through the Strait of Dover waxes and wanes in strength throughout the year 'in a sort of buffer relationship with the current from the north ... there exists a sort of see-saw conflict between the two.' (Carruthers, 1935; Russell, 1938)

The water entering the Channel from the west, formed partly of Atlantic and partly of coastal water, is often termed 'western water' in contrast to 'Channel water' from which it is chemically distinct by being richer in plant nutrients and of higher salinity (probably due to the upwelling of bottom water in the north of the Celtic Sea). As Russell (1939) puts it, the N. Sea can be regarded as a large bay 'containing water to which fresh water is added from the land and flowing out from the Baltic. At the same time the whole bay is open to an influx of oceanic water from the north, and through a bottle neck, the Strait of Dover, in the south. The surplus water flows out to the north.' It is in the middle regions of the N. Sea that the influence of the less saline waters from the Baltic, added to by fresh water from the rivers Elbe, Scheldt, Rhine and Weser, is most obvious and to some extent it separates the northern from the southern N. Sea. Even in its main features, and there are many complications which could be added, the water in the N. Sea has several origins. It might be thought that hydrographical conditions off our western coasts would be relatively simple, and indeed they consist mainly of coastal water to which is added Atlantic water from still farther west. But this is by no means the whole story. Leaving out details, it should be mentioned that off the western entrance to the Channel, over the continental shelf south of Ireland, there is a large cyclonic swirl which varies in position and extent.

Under the action of the prevailing westerly winds, more water is taken to the western entrance of the Channel than can enter it to pass through the Strait of Dover, and the surplus passes north and north west past the Scillies. As has been mentioned above, the amount of water entering the Channel from season to season and from year to year varies. The water in this swirl is the rich, so-called 'western water,' regarded by Southward (1961 and 1962) as 'N.W. water' in contrast to warmer water approaching the Channel entrance from further south, which he terms 'S.W. water.' It is a main contributor to the water of the Irish Sea, but along the western coast of Ireland at a depth of 600 metres or so a tongue of warm though dense (because of its high salinity) water, derived partly from deep water flowing out over the sill at the Strait of Gibraltar and partly from warm Atlantic water in the Bay of Biscay (Cooper, 1952), is deflected northwards, gradually approaching the surface along its course. The point at which it upwells to mix with oceanic and coastal waters varies from year to year but may be so far north that it carries its characteristic 'Lusitanian' plankton right up to the Faeroe Channel, far to the north of Scotland (Fraser, 1952, 1955).

XV
Regional Characteristics of Plankton

THE important idea that certain of the more obvious and readily identifiable members of the macroplankton can be used as indicators of different kinds of water masses, particularly those around our coasts, is largely due to Russell (1935) who has elaborated and summarized the relevant information in 1939 and 1952 (see also Fraser, 1962). Southward (1962) has re-stated the requirements of a good indicator species which are: 'that it should be common enough in all samples from an area; easy to pick out under low magnification; capable of being linked with some projected centre of origin or abundance from which it spreads or is dispersed, and it should not have too much facility for rapid reproduction.' It is true that many of the main water masses can be followed by their temperatures, salinity and other chemical peculiarities, but since each supports characteristic assemblages of planktonic organisms, these form, as it were, labels or tags which can be used by the planktonologist for the same purpose. Indeed, as Rae (1956) points out:

'There has been a tendency to presume that the gross physical and chemical parameters are the all-important ones. Temperature and the availability of organic nutrients have been singled out for particular attention possibly because of following closely the analogy of earlier conclusions on terrestrial ecology, or because of the limited selection of parameters available to the ecologist who is usually forced to work with observations taken by the oceanographer for his own very different objectives. However, there is now a growing weight of opinion that other less obvious factors play the prime role in determining many of the variations of marine populations both in space and time.'

Plankton assemblages, it seems, often form the surest means of recognizing waters of particular origin. So much is now known that Russell (1939) has been able to construct a list of the indicator species of all the main sea areas round the British Isles and the information given below is largely abstracted from his papers which should be consulted for further details. Various numbers of the *Hull Bulletins of Marine Ecology* contain much information on plankton indicators and Fraser (1965) has recently summarized the zooplankton species of the North Sea Area. For descriptive purposes we may conveniently divide the water masses around the British Isles into three main areas; (a) The North Sea, (b) The English Channel and waters off southern Ireland and (c) The Irish Sea and western coasts of the British Isles.

(a) The North Sea

Fraser (1965) has divided the North Sea into eight regions which correspond with different zooplankton communities although he emphasizes that the divisions are very variable and change not only with season but with depth. This distribution pattern is illustrated in figure 16 (on page 141) and the indicator species for the different areas are listed below.

Area 1. - North Atlantic Water

This contains a mixture of cold-water species entering from the Norwegian Sea or from areas of upwelling and of warmer-water species. Indicators of the cold water include the chaetognaths *Sagitta maxima*, *Eukrohnia hamata* and the copepods *Calanus hyperboreus*, *Metridia longa* and *Euchaeta norvegica*. Warm-water species include salps and doliolids, siphonophores especially *Physophora hydrostatica* and *Agalma elegans*; chaetognaths such as *Sagitta serratodentata* and *S. lyra*; copepods such as *Rhincalanus nasutus* and *Eucalanus elongatus*; and pteropods such as *Euclio pyramidata*.

Area 2. - Channel water in the southern North Sea

This water enters the southern North Sea through the Strait of Dover and so consists partly of Channel water proper and partly of waters with which it has mixed. Indicators of pure Channel water are *Biddulphia sinensis*, *Turritopsis* sp., *Gossea corynetes*, *Oithona nana*, mysids such as *Heteromysis* sp. *Mesopodopsis slabberi* and *Gasterosaccus Sanctus*. The chaetognath *Sagitta setosa*, the doliolid *Doliolum nationalis* and the polychaete *Greefia celox* are often abundant as well as larvae of *Lamellaria perspicua* in the spring and early summer.

Area 3. - Skagerrak water

Cold water forms such as the chaetognath *Eukrohnia hamata*, the copepods *Metridia longa* and *Calanus hyperboreus* and the euphausid *Meganyctiphanes*

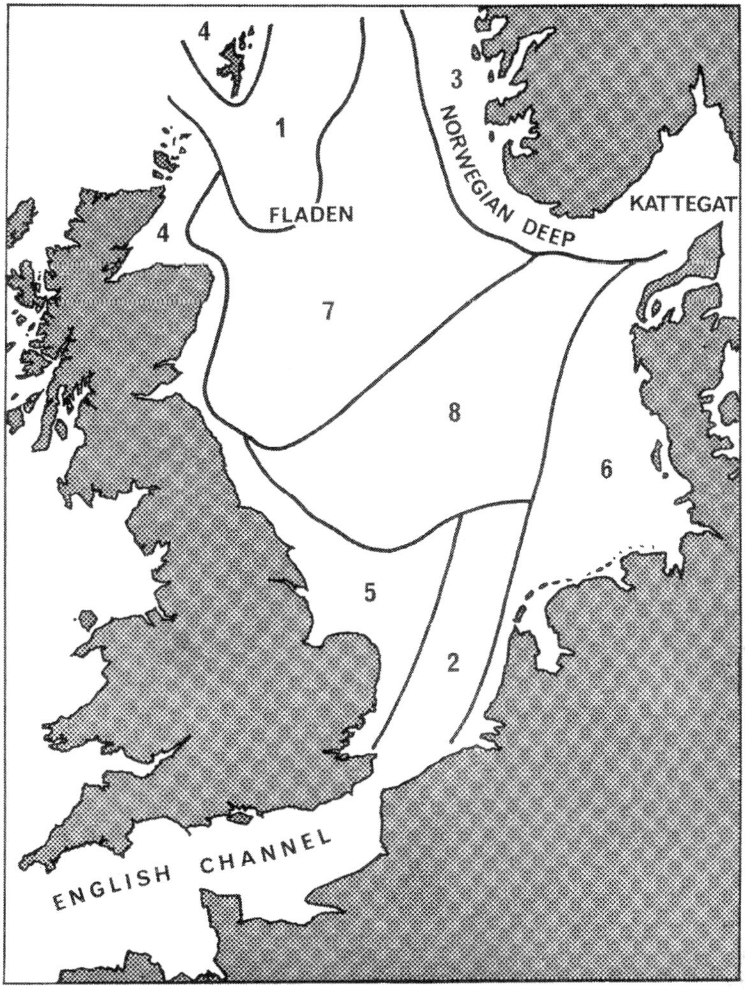

Figure 16. Map showing the distribution of indicator species in the North Sea (After Fraser, 1965)

norvegica occur at depth. Surface forms include *Sagitta elegans*, *Temora longicornis*, *Candacia norvegica* and *Euterpina acutifrons*.

Area 4. - Scottish coastal water

Much of the plankton of this area derives from mixed oceanic and coastal waters to the north and west of Scotland. This water is characterized by the medusae *Neoturris pileata*, and *Cosmetira pilosella*, the chaetognath *Sagitta elegans*, the pteropod *Spiratella retroversa* and the copepods *Euchaeta hebes*, *Temora longicornis*, *Metridia lucens* and *Centropages hamatus*.

Area 5. - Coastal water of Eastern England

Copepods are the main indicators of this area. Typical species include *Oithona nana*, *Labidocera wollastoni* and *Isias claviceps*, *Paracalanus parvus*, *Centropages hamatus* and *Euterpina acutifrons*. The diatom *Biddulphia regia* is also characteristic as well as the mysids *Mesopodopsis slabberi* and *Paramysis spiritus*.

Area 6. - Coastal water of the continent

Here *Sarsia tubulosa* is characteristic as well as *Noctiluca miliaris* and the copepods *Paracalanus parvus*, *Centropages hamatus*, *Acartia clausi*, *Euterpina acutifrons* and *Corycaeus anglicus*. The mysid *Gasterosaccus sanctus* is often found.

Area 7. - Northern North Sea water

Many species may be found in this area including both cold-water forms from area 1 and 3 as well as mixed oceanic and coastal forms from the north and west of the British Isles. Some of the more common species are the medusae *Neoturris pileata*, *Laodicea undulata* and *Cosmetira pilosella*, the siphonophore *Agalma elegans* and the polychaete *Tomopteris septentrionalis*. *Sagitta elegans* is the characteristic chaetognath and *Spiratella retroversa* as well as *Clione limacina* are often common. Typical crustaceans include the amphipod 'Themisto' sp., the euphausids *Meganyctiphanes norvegica* and *Thysanoessa inermis*, the mysid *Gasterosaccus spinifer* and the following copepods: *Candacia armata*, *Metridia lucens*, *Acartia clausi* and *Microsetella norvegica*.

Area 8. - Central North Sea

This water is characterized by *Sagitta setosa* and the following copepods, *Acartia clausi*, *Centropages typicus* and *Temora longicornis*. The coelenterates *Tima bairdi*, *Eutonina indicans* and *Cyanea capillata* are also typical.

(b) English Channel and waters off southern Ireland

Here can be recognized:

(i) Channel water characterized by a general paucity of plankton and with *Turritopsis* and *Sagitta setosa* as indicators.

(ii) Mixed oceanic and coastal water, which forms a large swirl off southern Ireland and may get carried into the channel - so called 'western-water' or 'N.W. water' of Southward (1962) indicated by *Laodicea undulata*, *Cosmetira pilosella*, *Aglantha digitale*, *Agalma elegans*, *Nanomia cara* and *Sagitta elegans*.

(iii) Warm oceanic water mainly from a south-westerly direction with indicators *Liriope tetraphylla*, *Muggiaea kochi*, *Euchaeta hebes* and various salps and doliolids.

(c) Irish Sea and western coasts of the British Isles

As mentioned, the Irish Sea derives most of its water from that lying outside the western entrance to the Channel, but it also receives water from several rivers. Its plankton is characterized by *Sagitta elegans* although *S. setosa* is found in the Bristol Channel and Morecambe Bay. *Sagitta elegans* is the dominant chaetognath along the western coasts of Ireland, whilst the Lusitanian plankton at deeper levels has already been noted. This, which may not reach the surface until N.W. Scottish waters, is characterized by an abundance of salps, doliolids, siphonophores and of exotic species typical of much warmer waters. Southward (1961 and 1962), using stramin nets and a modified Gulf III apparatus, has resurveyed the macroplankton of the Channel and its western approaches. In so doing, he has pointed out the importance of paying more attention to the north-south difference in the constitution of the plankton of this sea area (which roughly marks the boundary between many northern and southern species used as indicators). In redefining the indicator species of the water masses of the western approaches, Southward took account of whether a species was neritic or oceanic in this particular area since the on- or off-shore distribution may vary a good deal with latitude. He also stresses the fact that many species have temperature preferences. For these reasons he divides his indicator species for the Channel, and the sea off its western entrance, into north-western colder species and south-western warmer species.

North-western species include *Tomopteris helgolandica*, *Meganyctiphanes norvegica*, *Spiratella retroversa*, *Nanomia* sp. and *Thysanoëssa inermis*. These do not occur with any regularity in the Channel but are mainly deeper water species although present in the Fastnet area and Irish Sea. *Aglantha digitale* and larvae of *Luidia sarsi* are not abundant in the Fastnet area every year and do not enter the Irish Sea. Other northern species are *Sagitta elegans* and the herring, both of which are regularly taken in the western basin.

Southern species useful as indicators of this type of water are the oceanic *Sagitta serratodentata*, *Rhincalanus nasutus*, species of *Candacia* (other than *armata*) and species of *Salpa*. The pilchard must also be regarded as a southern species.

South-western species of fairly regular occurrence in the western approaches to the Channel include *Liriope tetraphylla* and species of *Doliolum*. Warm-water Biscay species are represented by *Muggiaea* sp., *Euchaeta hebes* and *Spiratella lesueuri*.

Western-basin species present all the year round, but commoner in the winter, are *Nyctiphanes couchi*, species of *Apherusa* and *Centropages typicus*. *Centropages hamatus* is a copepod typical of inshore waters of the eastern basin.

Plankton analysis

In most plankton studies the investigator will know, usually in detail, the area from which his samples have been drawn. Yet it sometimes happens that a student is given a sample and is then asked to make inferences about its place of origin. Usually only a very general answer can be given but inspection of the sample and reference to the previous sections will show the presence of indicator species of the main hydrographic areas around the British Isles. When we remember that the British Isles draw much of their water from the west and that as this water circulates to the north and east into the North Sea its indicator species gradually fade out, the hardier ones surviving longer than those more susceptible to changing conditions, it becomes possible to infer the distance at which the sample was taken from the centre of dispersion of the assemblage. Certain local conditions result in the presence of characteristic plankton assemblages; these deserve special mention and are described below.

(a) Inshore Plankton

A sample taken from near the shore would be expected to contain a mixture of deep and shallow water species to an extent which will vary with local conditions. In general offshore animals such as euphausids, chaetognaths other than *Sagitta elegans* and *S. setosa*, siphonophores (except when drifted inshore by winds) *Euchaeta* sp. will be rare or absent. The sample will be rich in organic debris. On the other hand, particularly in spring it will contain large numbers of larvae of intertidal invertebrates such as *Littorina littorea*, *L. neritoides*, acorn barnacles, *Nassarius* sp., *Trivia* sp., *Natica* sp., *Mytilus*, polychaetes (particularly nereids and spionids) *Carcinus maenas*, porcellanids and many others. Such organisms are mainly the larvae of local adults. Some of the holoplanktonic organisms may also be purely local inshore species such as shrimps, prawns, mysids, isopods, amphipods and cumaceans. Indeed, at night many sand and weed-dwelling crustaceans leave the sea floor and swim in the inshore waters where they form a special kind of temporary plankton (Colman and Segrove, 1955). Other small organisms, e.g. harpacticids, may get whirled up by the turbulence together with plant debris, polychaete bristles as well as cast cuticles of barnacles. In water over mud flats *Hydrobia ulvae* commonly occurs in the plankton collected at and around the time of high water. (Newell, 1962.)

(b) Estuarine Plankton

This may be recognized by the presence of many euryhaline species such as the mysids *Praunus flexuosus*, *Mesopodopsis slabberi*, *Gastrosaccus* sp., *Leptomysis gracilis* and *Neomysis integer*. Of the copepods, the following are likely to be represented: *Eurytemora hirundoides*, *Temora longicornis*, *Acartia clausi* and *Centropages hamatus*. Larvae of *Carcinus maenas* are often abundant and the isopods *Idotea linearis* and *I. balthica* often swarm in the surface waters. Larvae of the euryhaline prawn *Palaemonetes varians* and of *Porcellana longicornis* are likely to be met with. *Littorina littorea* eggs and larvae are seasonally abundant. *Oikopleura dioica* is the only common pelagic tunicate. Nevertheless, many offshore species will enter and flourish in the lower reaches of the larger estuaries, e.g. that of the Thames (Wells, 1938; Maghrabi, 1955), but estuarine plankton is rarely rich in species, one organism or a few tending to be dominant. As in other inshore areas, detritus of various kinds is usually abundant.

(c) Offshore Plankton

This will, in contrast, contain a greater proportion of oceanic species, e.g. siphonophores, salps, doliolids, chaetognaths other than *Sagitta setosa* and *S. elegans*, and *Euchaeta* sp. Larvae of intertidal animals are rare, e.g. veligers of *Nassarius incrassatus* and zoeas of *Porcellana platycheles* are confined to inshore plankton but those of the more widely distributed *N. reticulatus* and *P. longicornis* are found in both inshore and offshore plankton (Smith, 1953; Lebour, 1947). Offshore waters are almost free from detritus of all kinds.

(d) Western areas

Plankton collected some way off the western entrance to the Channel would be expected to contain numerous species of medusae including *Eutima gracilis*, *Aglantha digitale*, *Sarsia prolifera*, *Dipurena halterata*, *Cladonema radiatum* and *Zanclea costata* as well as many others which occur all round our coasts. A mixture of *Sagitta elegans*, *S. setosa* and *S. serratodentata* may be anticipated. Siphonophores such as *Muggiaea kochi*, *Agalma elegans*, *Nanomia bijuga*, *Velella* and *Physalia* may be found. The ctenophore *Bolinopsis* may on occasions be abundant. Alciopids, tomopterids and typhloscolecids and, in fact, the planktonic polychaetes in general are common only in western waters. *Calanus helgolandicus* tends to replace *C. finmarchicus* whilst *Rhincalunus nasutus* is sometimes common. *Euchaeta hebes* may be found and the cladocerans *Evadne* and *Podon* are often abundant.

Pilchard eggs often occur in vast numbers in the spring. Larvae of invertebrates more common in western areas are found at the appropriate season including the phyllosoma larva of *Palinurus*, nauplii of *Chthamalus stellatus*, larvae of *Pagurus prideauxi*, portunid larvae, larvae of *Porcellana platycheles* larvae of *Maia squinado* as well as various characteristic veligers of gastropods. A somewhat similar plankton occurs in the Irish Sea.

XVI
Seasonal Characteristics of Plankton

AS IS well known, the nature of the plankton from any locality in temperate waters varies greatly from season to season. In British waters there is a great outburst of phytoplankton in the spring, made particularly obvious in samples by the numerous diatoms. This reaches a maximum in most areas in March and April, and is shortly followed by an increase in the zooplankton, many of whose members graze on the rich sea pastures so provided. Indeed, this grazing may be so intense that the phytoplankton is reduced before the plant nutrients have fallen to a level below which phytoplankton cannot be abundant. Then, in early summer, when a thermocline is established, phytoplankton remains sparse, because the phosphates and nitrates of the euphotic zone are insufficient for much production of plant materials until the autumn when the thermocline breaks down and, aided by the equinoctial gales, the plant nutrients are renewed by the mixing of the surface water with the deeper waters that have stored the nutrients. Daylight and temperature in early autumn are such that a second phytoplankton outburst then occurs, but it is of a brief duration and throughout the winter the density of the phytoplankton is at a low level. Moreover, the species of diatoms in the spring are different, as a rule, from those in the autumn outburst.

These remarks apply only to fairly deep water areas. In others, where the water is shallow and there is considerable turbulence from wave action, a thermocline is never established, there is a constant supply of plant nutrients, and the phytoplankton may be abundant throughout the summer. This applies to some estuaries, for example that of the Thames. Nevertheless, the finding of many diatoms in a sample indicates that it has been taken either in spring or autumn as a rule. Such evidence must be taken in conjunction with other indications. For example, most marine benthonic invertebrates breed in the spring and the finding of echinoderm larvae is a fairly sure indication of the sample's spring origin. The same is true for most polychaete larvae, for the nauplii of *Balanus balanoides*, for cyphonautes and for many others. Crustacean larvae in general are somewhat later in reaching their maximum and should, perhaps, be considered as summer members of the meroplankton. Many mollusc larvae reach their maximum in late summer and early autumn.

Of course the number of such small planktonts as diatoms in any sample depends, not only on the season, but on the mesh of net employed. Yet even in coarse net samples a large number of diatoms are usually retained in spring hauls. In short, the presence of a variety of diatoms and invertebrate larvae may be taken as indicating a spring origin for a sample.

Early autumn samples may contain many diatoms but larvae, apart from those of molluscs, are few in variety and number. Winter plankton is usually sparser in all respects but is certainly deficient in larvae, apart from those of certain species, and phytoplankton.

Seasonal variation in the abundance of indicator species in the western approaches of the Channel is discussed by Southward (1962). The north-western species, with a centre of abundance in the Fastnet region, are restricted to this region from December to March but extend southwards across the Channel from May to July after which they retreat northwards. South-western species (e.g. *Liriope* and *Euchaeta*) tend to be restricted to the area south-west of Ushant in May or June, but their centre of abundance moves northwards in July only to go back again in February. These movements of indicator species are probably related to the extensions and contractions of the anti-clockwise swirl west of the Channel entrance.

Southward (1963) also discusses the reasons underlying changes in the nature of the plankton which have occurred in the last fifty years or so in the western Channel. He concludes that these are due more to rising temperature than to changes in the fertility of the water, and may be quite unrelated to variations in the amount of Atlantic water flowing into the Channel.

Appendix I
KEY TO COMMON COPEPODS

To distinguish adults from immature calanoids (See also p. 56)

Females - all females have a swollen genital segment and spermothecae are always present. There is a pair of small sacs in the first abdominal segment visible as white spots by refracted light and as dark spots by transmitted light.

Males - all males except those in categories A and B (below) have hinged antennules, usually on the right side. In males of class A, antennules are fringed with sensory papillae for about half their length. Males are less common than females.

Approximate identification

Four genera of copepod can be distinguished from other calanoids without the aid of a microscope. These are *Rhincalanus*, *Eucalanus*, *Euchaeta* and *Calanus finmarchicus*. All are large species from 2-4mm in length and can be distinguished from one another and from other genera by the use of the keys below.

Key to Major Categories

(This is not a taxonomic classification but an arbitrary way of identifying them.)

A. Body parallel-sided and sausage-shaped. Right and left antennules alike in the male (**26**.1-6, **27**.3)

B. Body terminating in an acute rostrum. Large size. Right and left antennules alike in the male (**26**.7)

Body broader at the anterior end than at the posterior end.

C. Short domed body (**27**.1)

D. Body broader at the centre than at either end (**28**.5, 6; **29**.1, 3)

E. Body with marked shoulders and a rather truncate anterior end. Posterior end terminating in two spine-like processes (**27**.4, **28**.1)

F. Body with a hastate anterior end. Posterior end terminating in two spine-like processes (**28**.2, 3)

G. Body as wide at back as at front (**29**.5)

CATEGORY A

Body parallel-sided and sausage-shaped. Right and left antennules alike in the male.

1. *5 segments behind head.* Large animal 4mm. long. Two long plumose setae at end of the antennules. *Calanus finmarchicus* (**26**.1)
2. *4 segments behind head.*
   (a) Female has 5th leg and may carry egg-sacs *Metridia* (**27**.3)
   (b) Large with a pointed head *Rhincalanus* (**26**.3)
   (c) Large with an angulated head *Eucalanus* (**26**.2)
3. *3 segments behind head.*
   (a) Urosome half length of metasome. Four pairs of swimming legs. Small copepod with a very small urosome *Pseudocalanus* (**26**.5)
   (b) Urosome a third the length of the metasome.
      (i) Female has small 5th leg. Body nearly three times as long as broad *Paracalanus* (**26**.4)
      (ii) Female has no 5th leg. Body twice as long as broad *Microcalanus* (**26**.6)

CATEGORY B

Body terminating in an acute rostrum. Large size. Right and left antennules alike in the male.

Euchaeta (**26**.7)

A very large copepod. Brightly coloured. Very projecting maxillipedes. Female may carry egg-sacs which are usually bright blue. *One of the setae on the caudal furcae is as long as the body.* Compound bristles on tail.

CATEGORY C

Body broader at the anterior end than at the posterior end.

Temora (**27**.1)

The caudal furcae are long and narrow.

CATEGORY D

Body broader at the centre than at either end.

1. *4 segments behind head.*

(a) Urosome nearly as long as metasome. A small transparent copepod with setae on the last segment of the urosome *Oithona* (**29**.1, 2)

(b) Urosome about half the length of the metasome. Prominent black eye. Setae on caudal rami spread out in fan. Antennae bear long sparse bristles *Acartia* (**28**.5-7)

2. *3 segments behind head.*

A small transparent copepod with red eyes *Cyclopina* (**29**.3)

CATEGORY E

Body with marked shoulders and a rather truncate anterior end. Posterior end terminating in two spine-like processes

1. *5 segments behind head.*

End of metasome produced into points, asymmetrical in males. Front of head pointed. Small fat body. Narrow front end of cephalothorax. Long antennae. Last thoracic segment has a point on each side *Centropages* (**27**.4, 5)

2. *4 segments behind head.*

End of metasome angular with pointed ends. Female with small 5th leg. Male and female with projections on genital segments. Front of head flat *Candacia* (**28**.1)

CATEGORY F

Body with a hastate anterior end. Posterior end terminating in two spine-like processes.

1. *5 segments behind head.*

Asymmetrical caudal furcae in female. Ventral eye in male, greenish-blue in colour. Rostrum has two strong hooks. Last thoracic segment has two hooks *Anomalocera* (**28**.2)

2. Much swollen genital segment in the female. Male has prominent lens on the dorsal side of the head *Labidocera* (**28**.3)

CATEGORY G

Body as wide at back as at front.

A pair of large crystalline 'eyes' anteriorly *Corycaeus anglicus* (**29**.5)

Appendix 2
KEY TO CRUSTACEAN LARVAE

Key to Major Categories

A. First three pairs of head appendages setose, other appendages absent or rudimentary NAUPLIUS (**40**)

B. Setose expodites on some or all of the thoracic appendages. Pleopods absent or rudimentary ZOEA (**41**-**43**)

MEGALOPA stages often abundant, are readily recognizable by their setose pleopods and can be identified by reference to the plates (**44**-**47**)

CATEGORY A. *NAUPLIUS*

1. (i) Fronto-lateral spines present 2
   (ii) Fronto-lateral spines absent 5
2. (i) Triangular outline, long posterior spine 3
   (ii) Rounded outline, short posterior spine, rounded distal end to labrum *Chthamalus stellatus* (**40**.7)
3. (i) Postero-lateral spines present
   (ii) Postero-lateral spines absent, rounded tip to the labrum; long fronto-lateral spines *Verruca stroemia* (**40**.1)
4. (i) Tip of labrum truncated *Balanus balanoides* (**40**.4)
   (ii) Tip of labrum trilobed with large median lobe projecting posteriorly *Elminius modestus* (**40**.6)
5. (i) Appendages with distinct segments in early stages. No yolk in body COPEPODA

(See Ogilvie, 1953, and Lovegrove 1956.
Fiches d'Identification. Sheets 50 & 63).

(ii) Appendages with no distinct segments in early stages. No masticatory hooks on the antenna and mandible. Body full of yolk. A transitory larva passing through all stages within twenty-four hours PENAEIDEA

CATEGORY B. *ZOEA*

1. (i) Carapace dorso-ventrally flattened, rounded in outline, and transparent: a phyllosoma *Palinurus* (**42**.4)
   (ii) Region between eyes and mouth extremely elongated: a trachelifer *Jaxea* (**43**.6)
   (iii) Carapace spherical dorsal and rostral spines usually present. When present, rostral spine points ventrally. (BRACHYURA **46**) 2
   (iv) Length of carapace considerably greater than breadth or depth, rostrum pointing forwards 4
2. (i) Rostral, dorsal and lateral spines present 3
   (ii) Rostral and lateral spines absent *Inachus dorsettensis* (**46**.6)
   (iii) Rostral and dorsal spines absent, lateral spines minute *Ebalia tuberosa* (**47**.4)
   (iv) Lateral spines absent, rostral and dorsal spines present *Carcinus maenas* (**46**.2)
3. (i) Dorsal spine long. Rostral spine long and straight; lateral spines short. Black and orange markings on carapace *Portunus puber* (**46**.1)
   (ii) Short curved dorsal spine, rostral spine short and straight. Lateral spines short. A very small zoea *Pilumnus hirtellus* (**46**.3)
   (iii) Both dorsal and rostral spines curved. Rostral spine distinctly shorter than dorsal spine. Lateral spines longer than in (ii). Greenish with yellow and black chromatophores *Maia squinado* (**46**.4)
4. (i) Postero-lateral margins of carapace produced into a spine on each side, or denticulate, or with both spines and denticles. (ANOMURA) 5
   (ii) Postero-lateral margins of carapace not produced into spines and not denticulate 14
5. (i) Rostrum extremely elongated: longer than the body. Postero-lateral spines extending beyond the telson. (PORCELLANIDAE) 6
   (ii) Rostrum shorter than body. Postero-lateral spines not extending beyond the telson 7
6. (i) Posterior spines half the length of the rostral spine *Porcellana platycheles* (**43**.1)
   (ii) Posterior spines one third the length of the rostral spine *P. longicornis* (**43**.3)

7. (i) Margins of spine denticulate. (GALATHEIDAE) 8

(ii) Margins of spine smooth. Eyes longer antero-posteriorly than the width of the abdomen. (PAGURIDAE) 10

8. (i) Rostral spine long; antennal scale with spiny tip. Paired dorsal spines on all abdominal segments, fourth and fifth have lateral spines also *Munida bamffica* (**42**.2)

(ii) Rostral spine and antennal scale smaller than in (i). No dorsal spines on posterior margin of abdominal segments 9

9. (i) Long lateral spines on abdominal segments four and five. Second to fifth abdominal segments serrated on posterior margin *Galathea strigosa* (**43**.7)

(ii) Short lateral spines on abdominal segements four and five *G. squamifera* (**42**.6)

(iii) Lateral spines present only on fifth abdominal segment *G. dispersa* (**42**.5)

10. (i) Antennal scale at least six times as long as broad 11

(ii) Antennal scale only four times as long as broad 12

11. (i) Large lateral spines on fifth abdominal segment; rostrum reaches the tip of the antennal scale *Pagurus pubescens* (**44**.3)

(ii) Fifth abdominal segment bears small lateral spines. Rostrum projects beyond spine on antennal scale. Longest spine on telson more than half greatest width of telson *P. bernhardus* (**44**.2)

12. (i) Very small mid dorsal spine on the posterior border of the sixth abdominal segment in late zoeas. Paired yellow chromatophores on abdominal segment 5. Orange chromatophores on each side of carapace. 4 pairs of pleopods in the final zoea *P. prideauxi* (**44**.5)

(ii) No mid dorsal spine on posterior margin of sixth abdominal segment of late zoeas 13

13. (i) A pair of prominent lateral spines on fifth abdominal segment. Longest telson spine less than half the greatest width of telson. Paired yellow chromatophores in early zoeas on the carapace and on the fourth and fifth abdominal segments. 4 pairs of pleopods in the final zoea *P. cuanensis*

(ii) Short lateral spines on fifth abdominal segment. 2 pairs of pleopods in the final zoea *Anapagurus hyndmanni* (**44**.4)

(iii) No lateral spines on fifth abdominal segment. 3 pairs of pleopods in the final zoea *A. laevis* (**44**.1)

14. (i) Antennal exopod segmented throughout its length. Telson with two cylindrical rami Early zoeas of PENAEIDEA (**47**.2)

(ii) Antennal exopod unsegmented, or segmented at distal end only. Telson flat or with tapering rami 15

15. (i) Telson ending in two sharp prongs or legs 1-3 chelate or with both these characters. Setose exopods on maxillipeds and legs 1-5 in all stages of most species — Last zoeas of PENAEIDEA

(ii) Telson not terminating in two sharp prongs. Leg 3 never chelate. Setose exopods confined to maxillipeds in early stages. Usually developed on some or all legs in later zoeas. (CARIDEA) — 16

16. (i) One or more of the thoracic endopods is oar-shaped — HIPPOLYTIDAE (*Caridion* sp.)

(ii) None of the thoracic endopods oar-shaped — 17

17. (i) Fifth leg much longer than the fourth. No exopod on leg 5. Endopod of maxilliped I small and unsegmented — ALPHEIDAE (**47**.6)

(ii) Fifth leg equal in length or shorter than leg 4. No exopod on leg 5. Endopod of maxilliped I well developed — 18

18. (i) No supra-orbital spine. Eyestalks hemispherical, nearly meeting in the mid-line. Antennules touching at base. Rostrum and telson may be broad. (CRANGONIDAE) — 19

(ii) Supra-orbital spine present in all except stage I zoeas. Eye-stalks cylindrical and often widely separated at the base. Rostrum tapering from base, telson may be narrow — 20

19. (i) Large dorsal spine on posterior margin of third abdominal segment. Small lateral spines on fifth abdominal segment. Antennal scale unjointed — *Crangon vulgaris* (**41**.7)

(ii) No dorsal spine on the posterior margin of the third abdominal segment. A pair of enormous lateral spines on fifth abdominal segment — *C. allmani* (**42**.3)

(iii) Two spines on the posterior margin of the third and fourth abdominal segments Peduncles of the antennules may be twice the length of the rostrum. Dark in colour. Much smaller than (i) and (ii) — *Philocheras fasciatus* (**42**.1)

20. (i) Base of antennules separated by more than the width of one of them. Eyestalks usually tapering. Peduncles of antennules slender and curved — 22

(ii) Base of antennules separated by not more than the width of one of them. Eyestalks cylindrical. Peduncles of antennules stout and nearly straight. (HIPPOLYTIDAE) — 21

21. (i) A pair of small lateral spines on the fifth abdominal segment. Three spines on the ventro-lateral margin of the carapace — *Hippolyte varians* (**41**.8)

(ii) A pair of large lateral spines on the fifth abdominal segment. Five lateral spines on the ventro-lateral margin of the carapace — *H. prideauxiana*

22. (i) Rostrum usually long and is toothed in late zoeas. Antennal scale usually segmented at tip in early zoeas. Exopods on legs 1-3 in late stage zoeas. (PANDALIDAE) 23

(ii) Rostrum small or absent, never toothed. Antennal scale never segmented. Exopods on legs 1-4 in late zoeas PROCESSIDAE (**41**.6)

23. (i) First zoea 3-4mm. long with a full complement of limb rudiments. Long rostrum projecting in front of the eyes *Pandalus montagui* (**41**.2)

(ii) First zoea 1.5-2.0mm. long with no limb rudiments. Short rostrum *Pandalina brevirostris* (**41**.3)

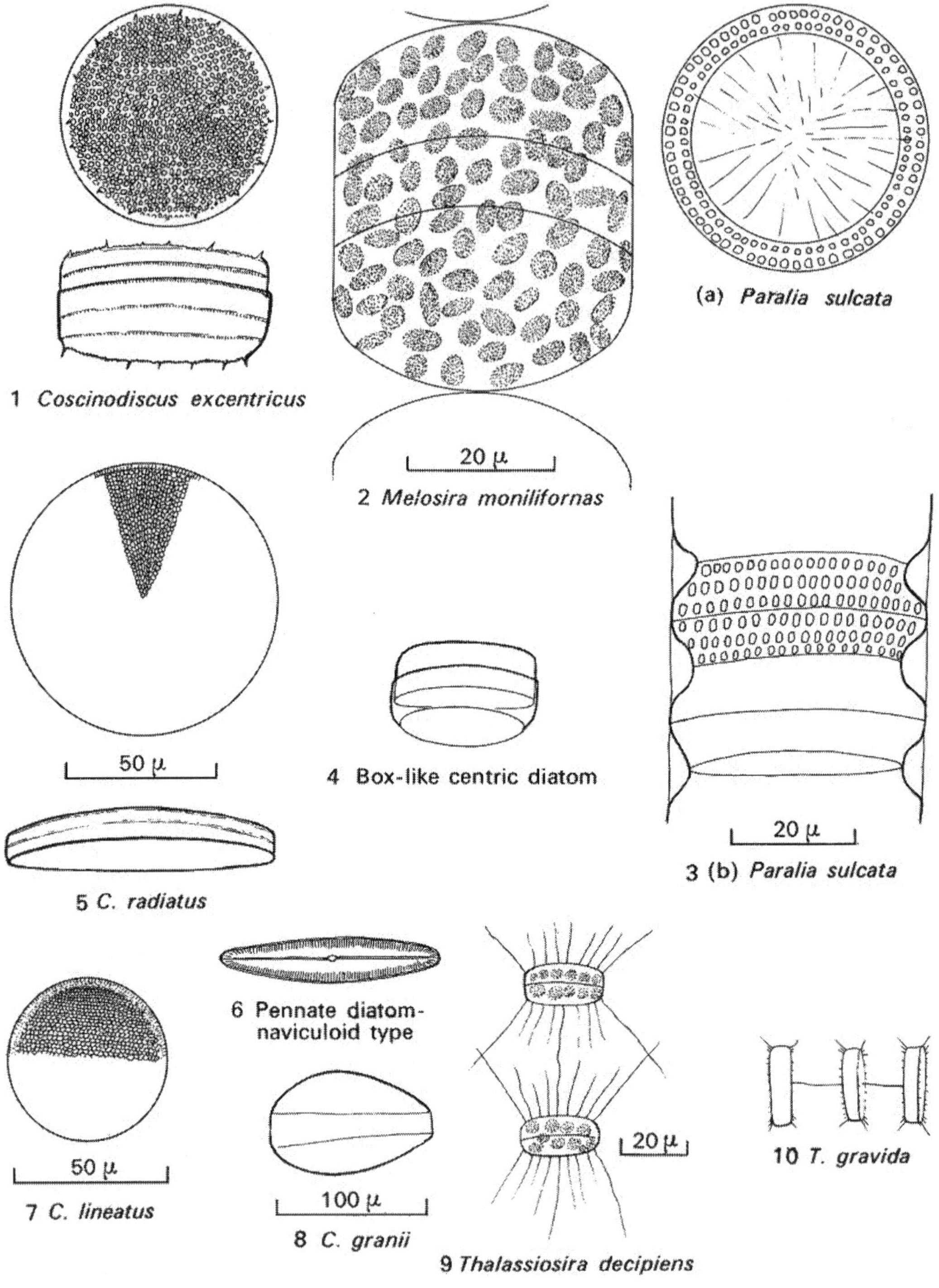

Plate 1. Centric diatoms

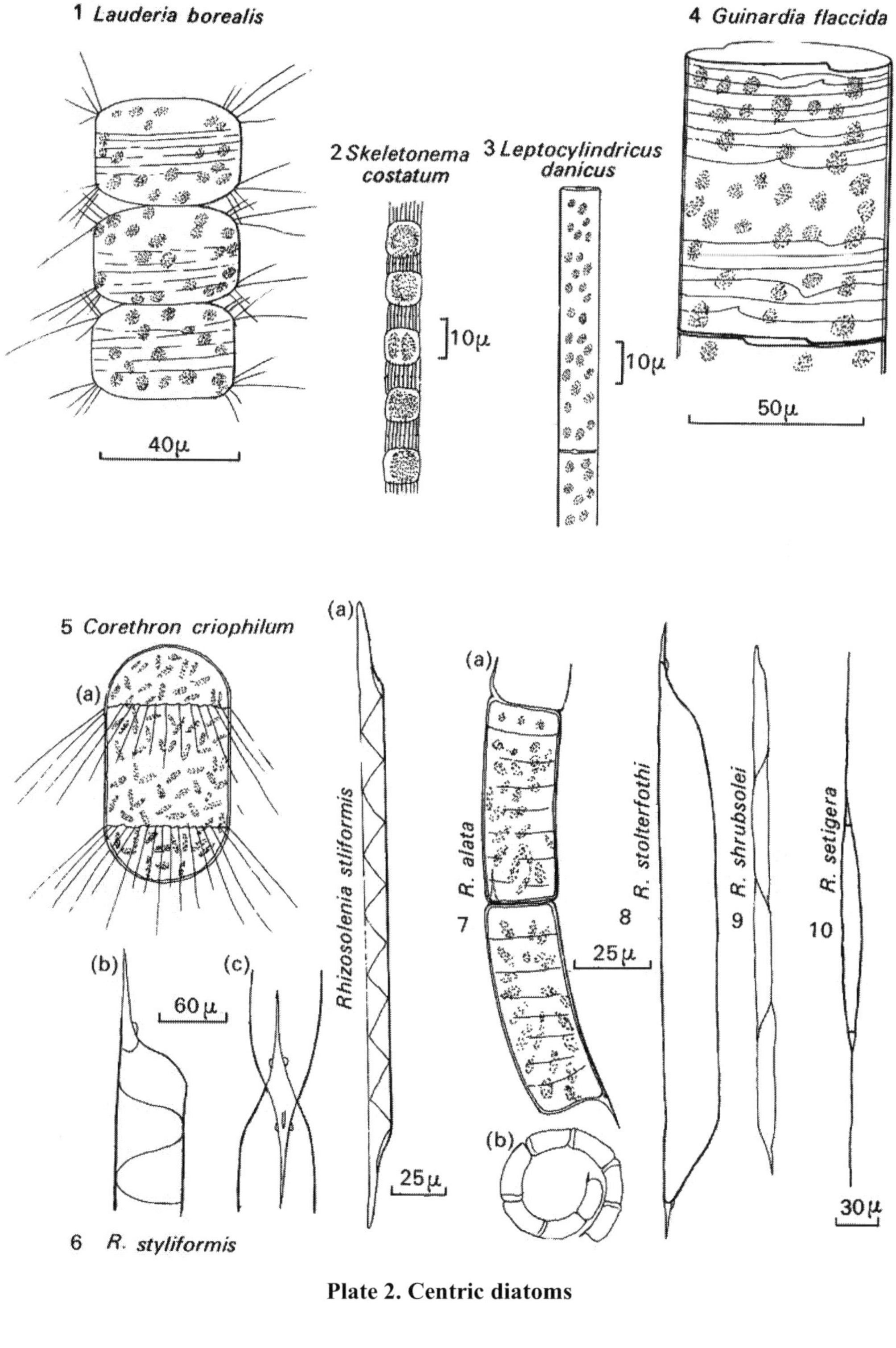

Plate 2. Centric diatoms

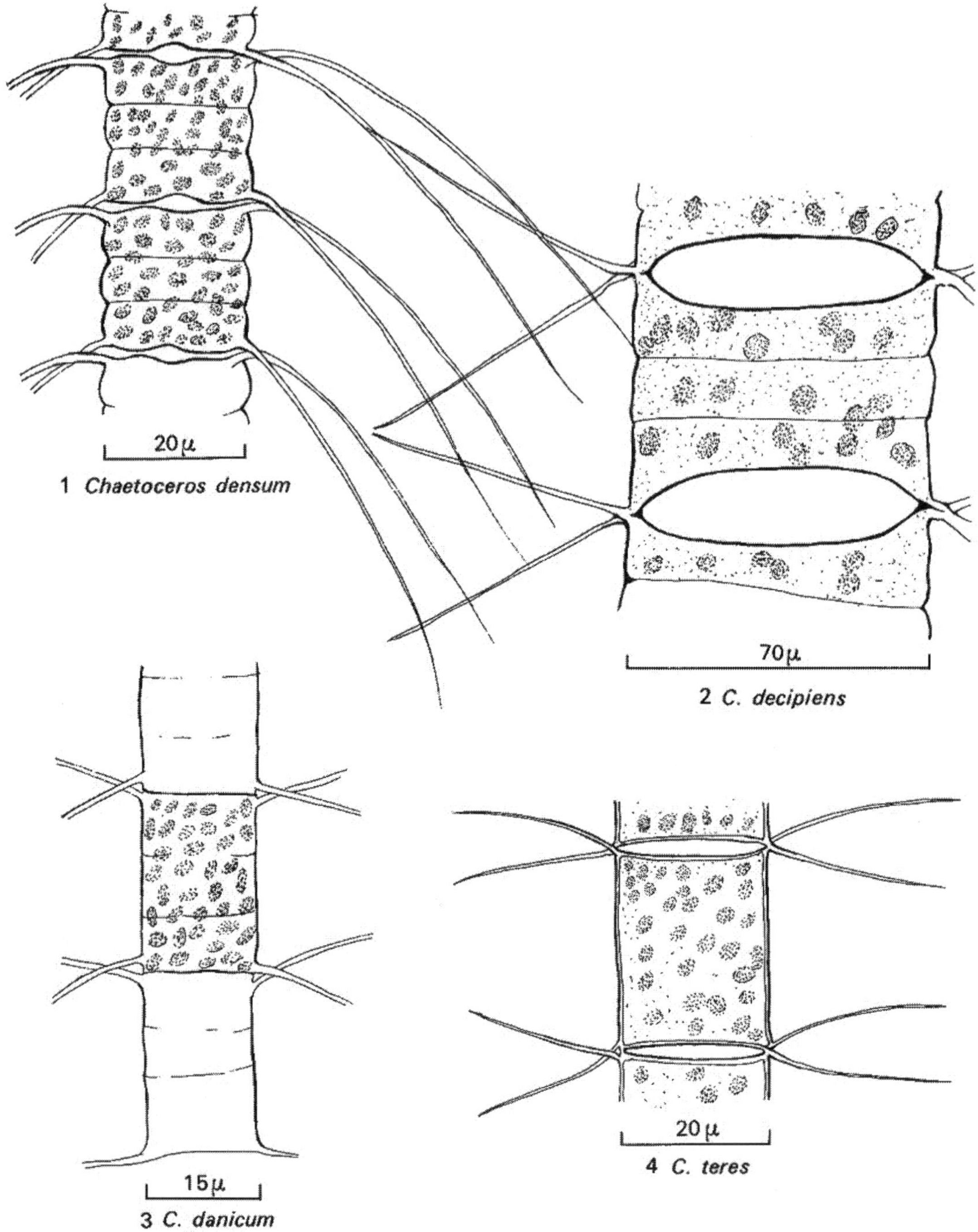

Plate 3. Centric diatoms

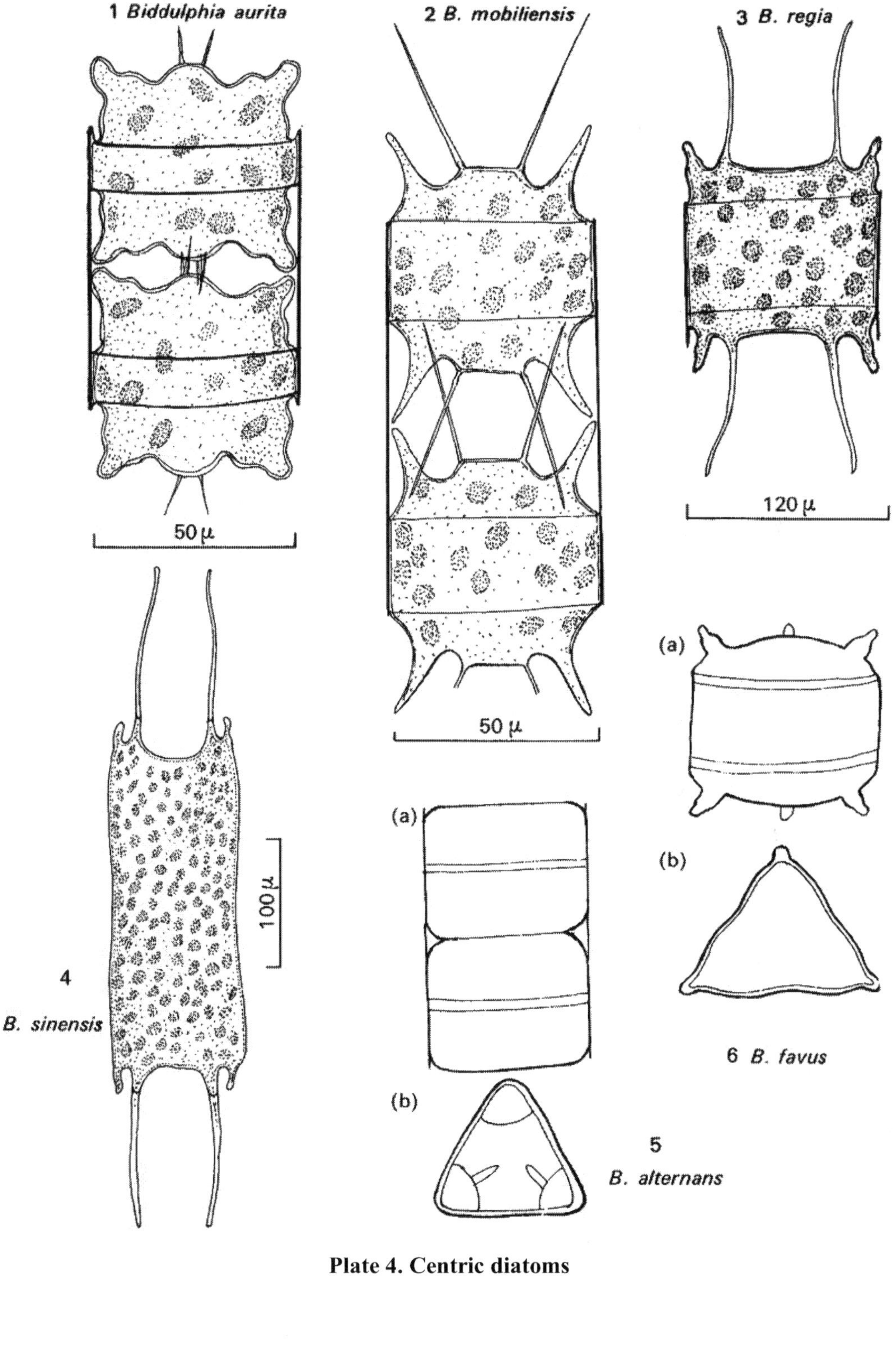

Plate 4. Centric diatoms

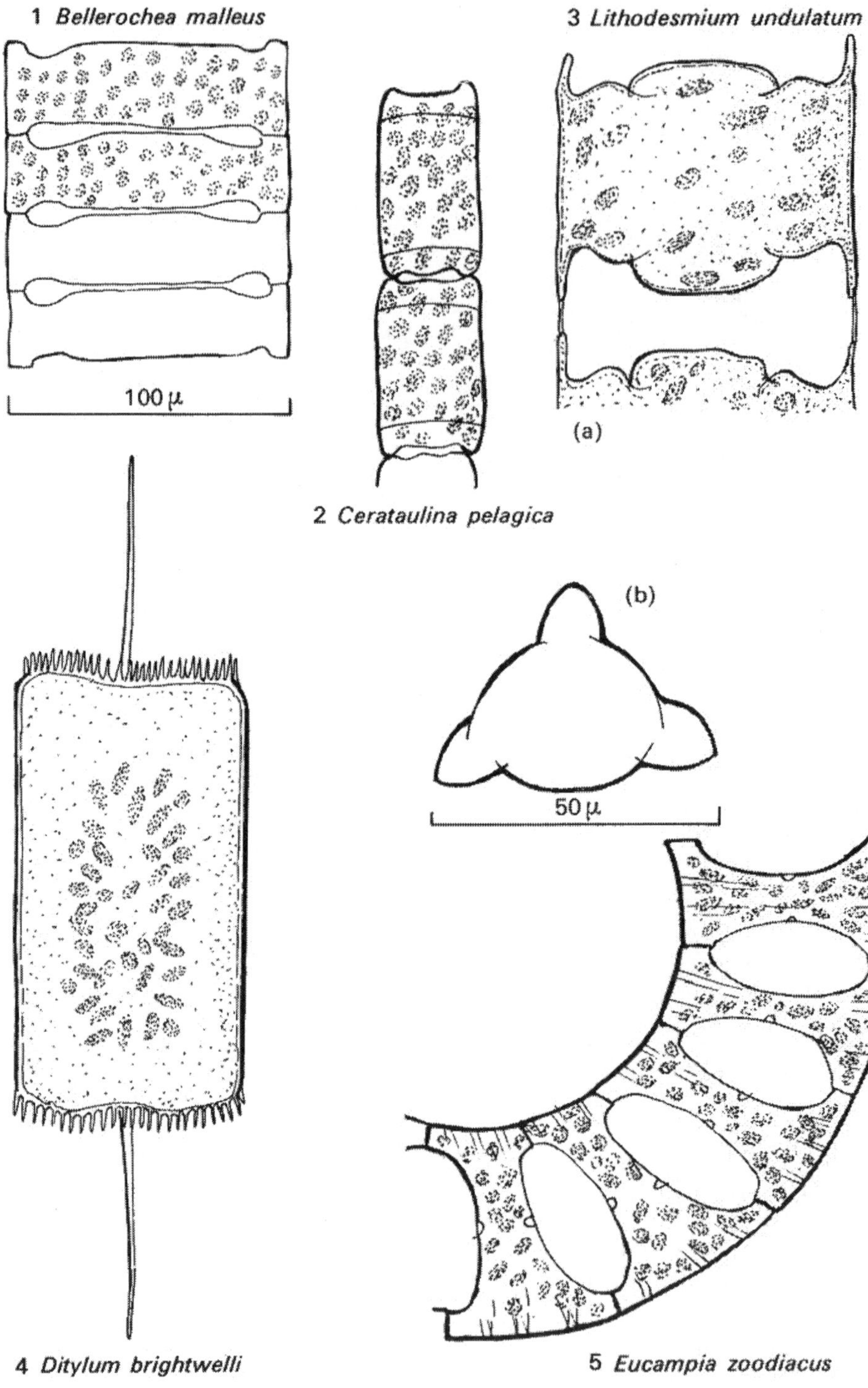

Plate 5. Centric diatoms

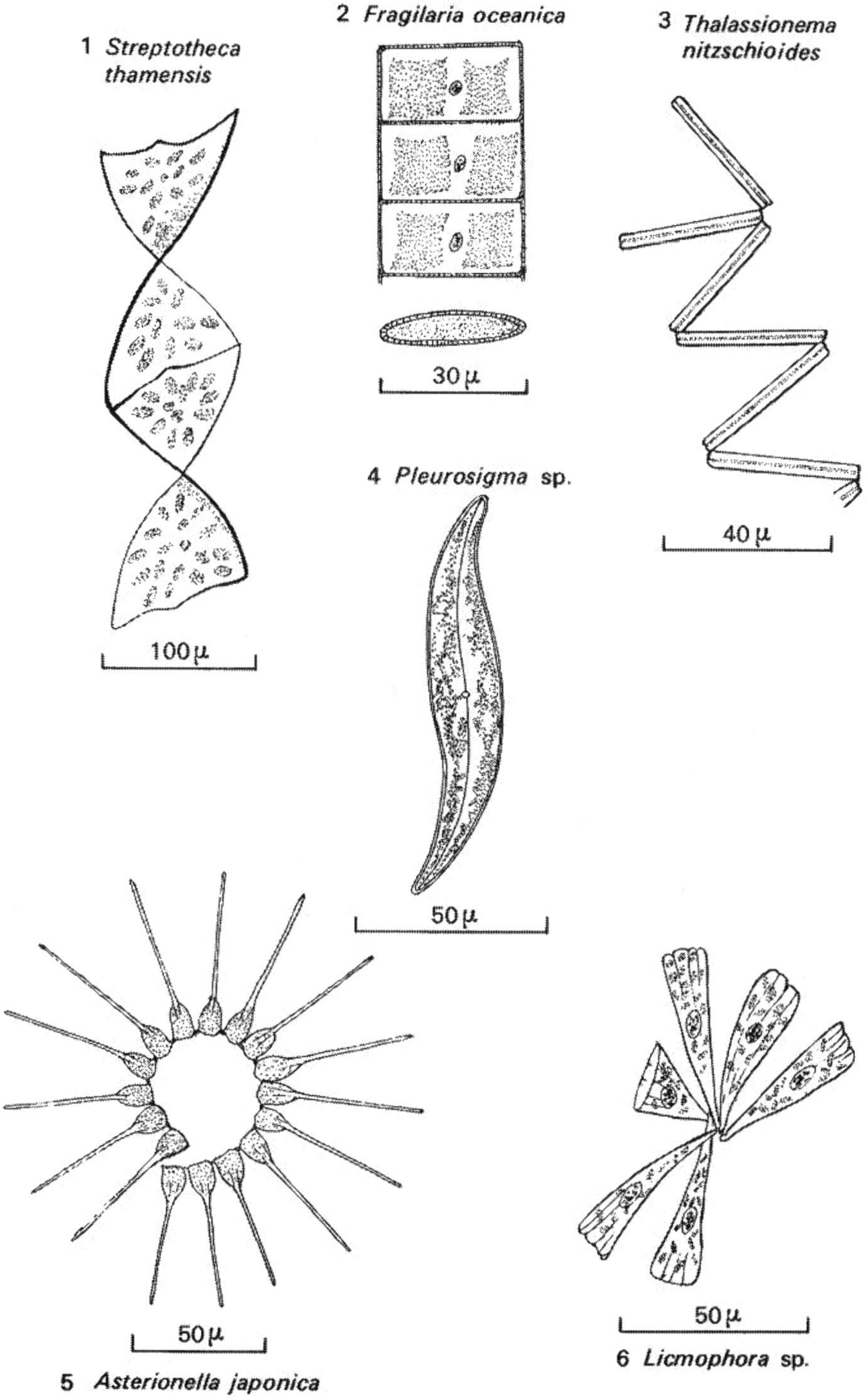

Plate 6. Centric and Pennate diatoms

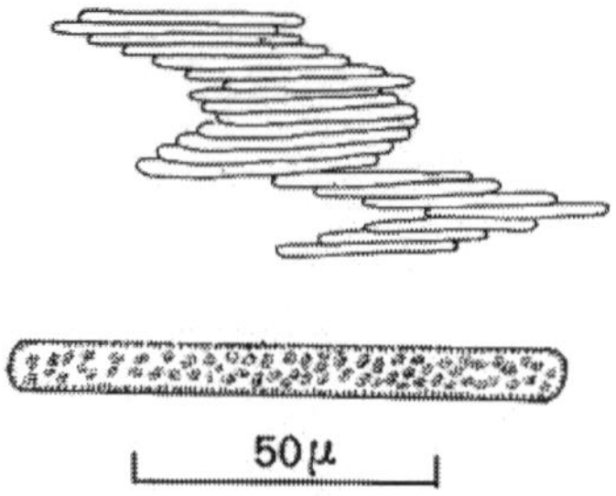

1 *Bacillaria paradoxa*

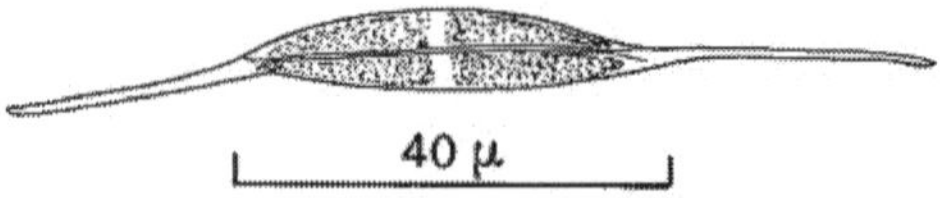

2 *Nitzschia closterium*

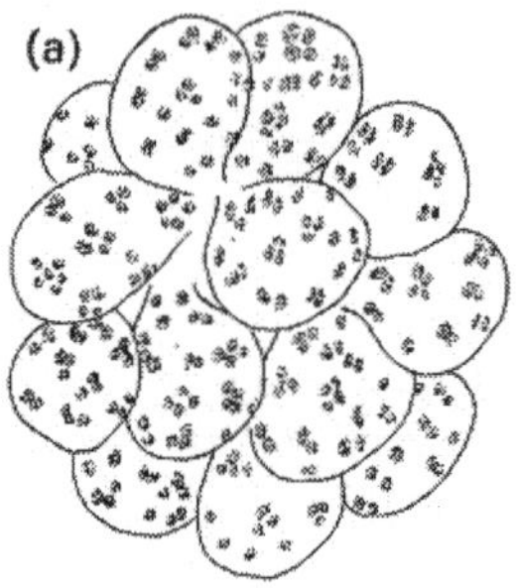

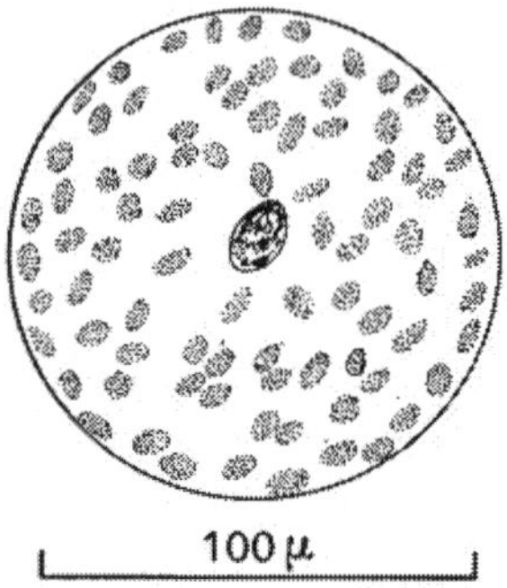

4 *Halosphaera viridis*

3 *Phaeocystis*

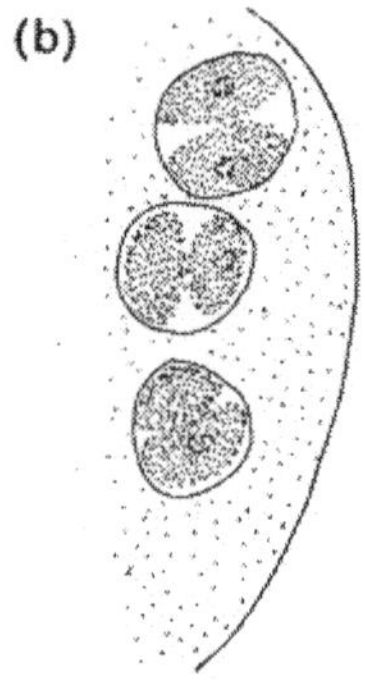

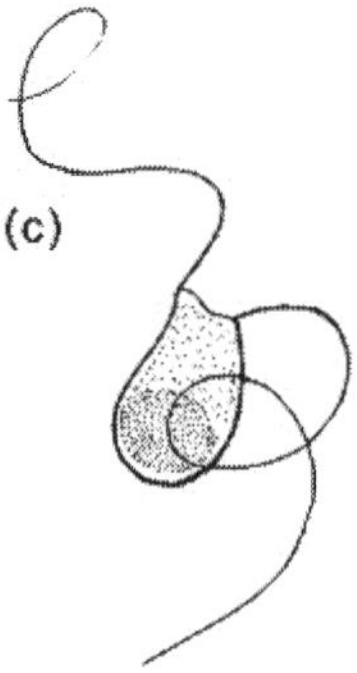

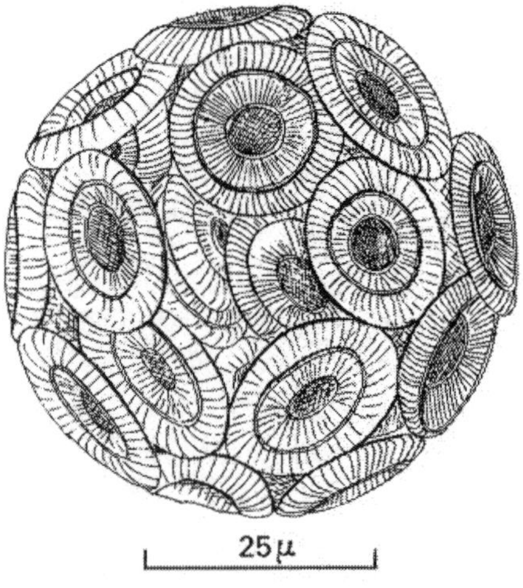

5 A coccosphere

Plate 7. Pennate diatoms, *Halosphaera*, *Phaeocystis* and Coccoliths

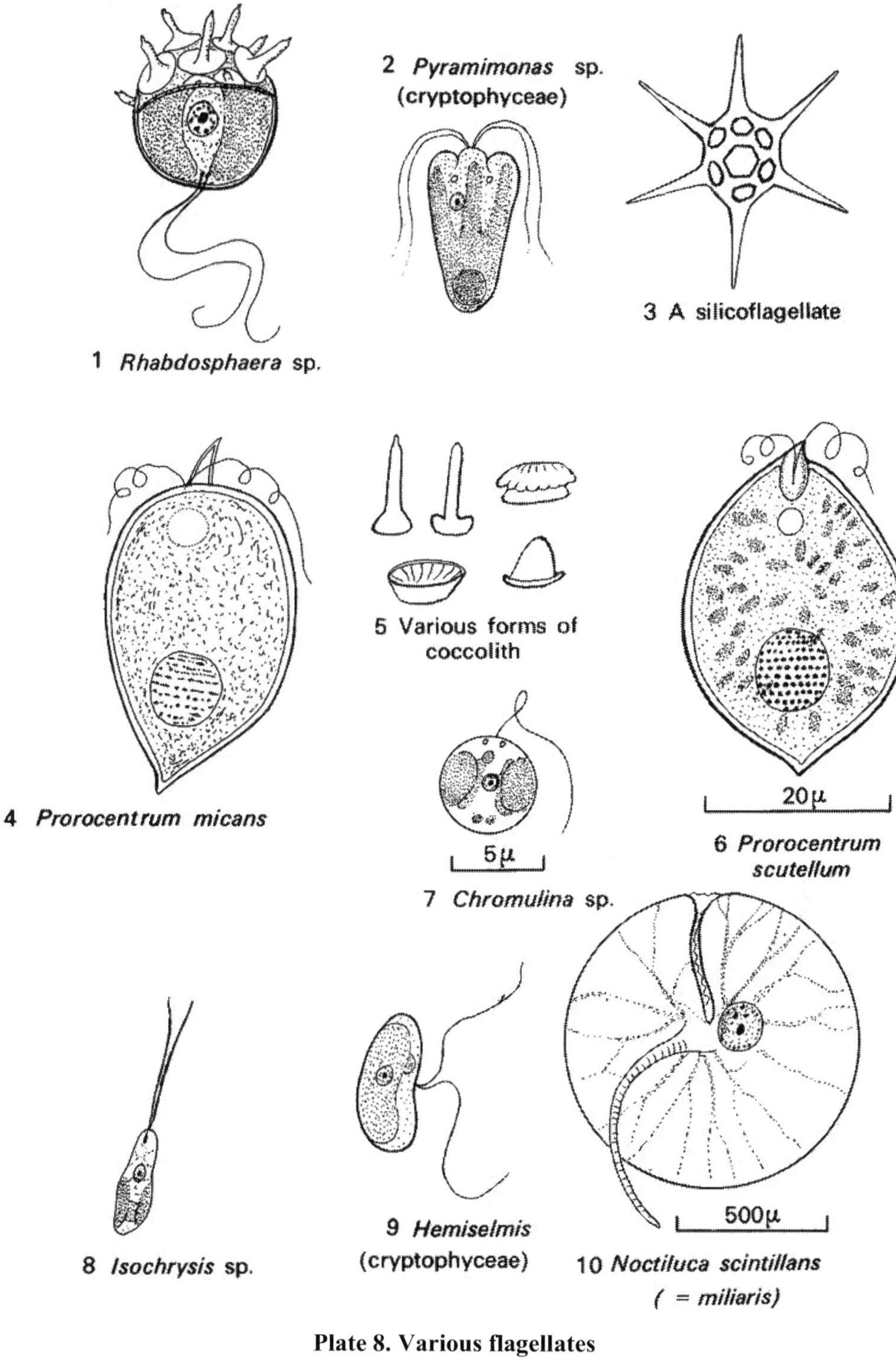

Plate 8. Various flagellates

1 *Gymnodinium(=Pyrocystis) lunula*
(not to scale)

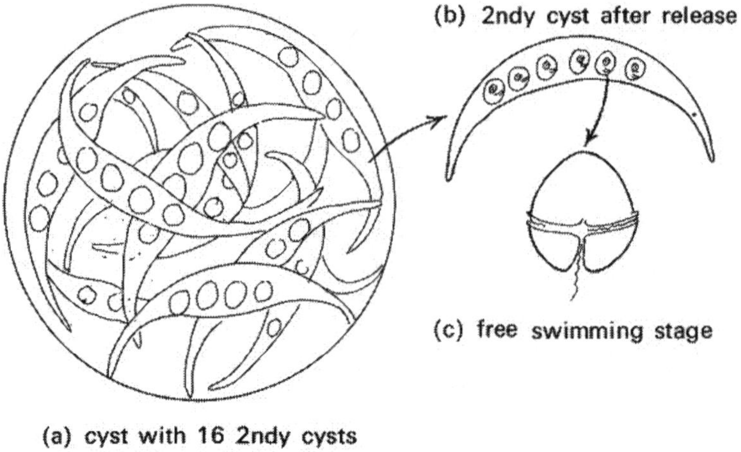

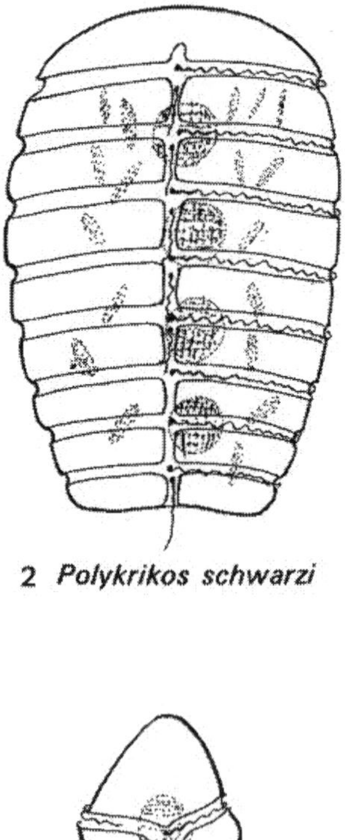

2 *Polykrikos schwarzi*

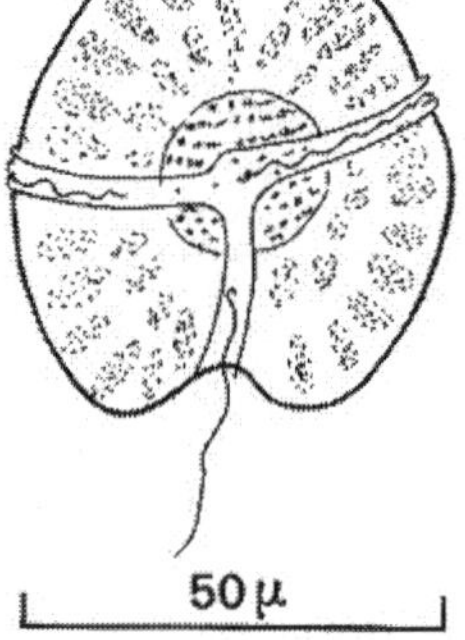

3 *Gymnodinium splendens*

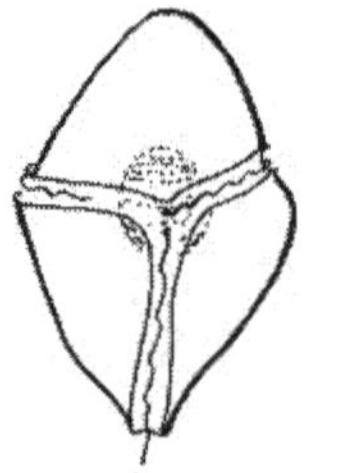

4 *G. rhomboides*

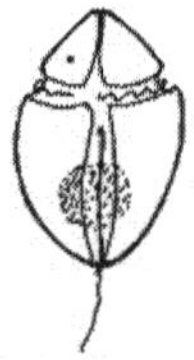

5 *Amphidinium crassum*

Plate 9. Dinoflagellates

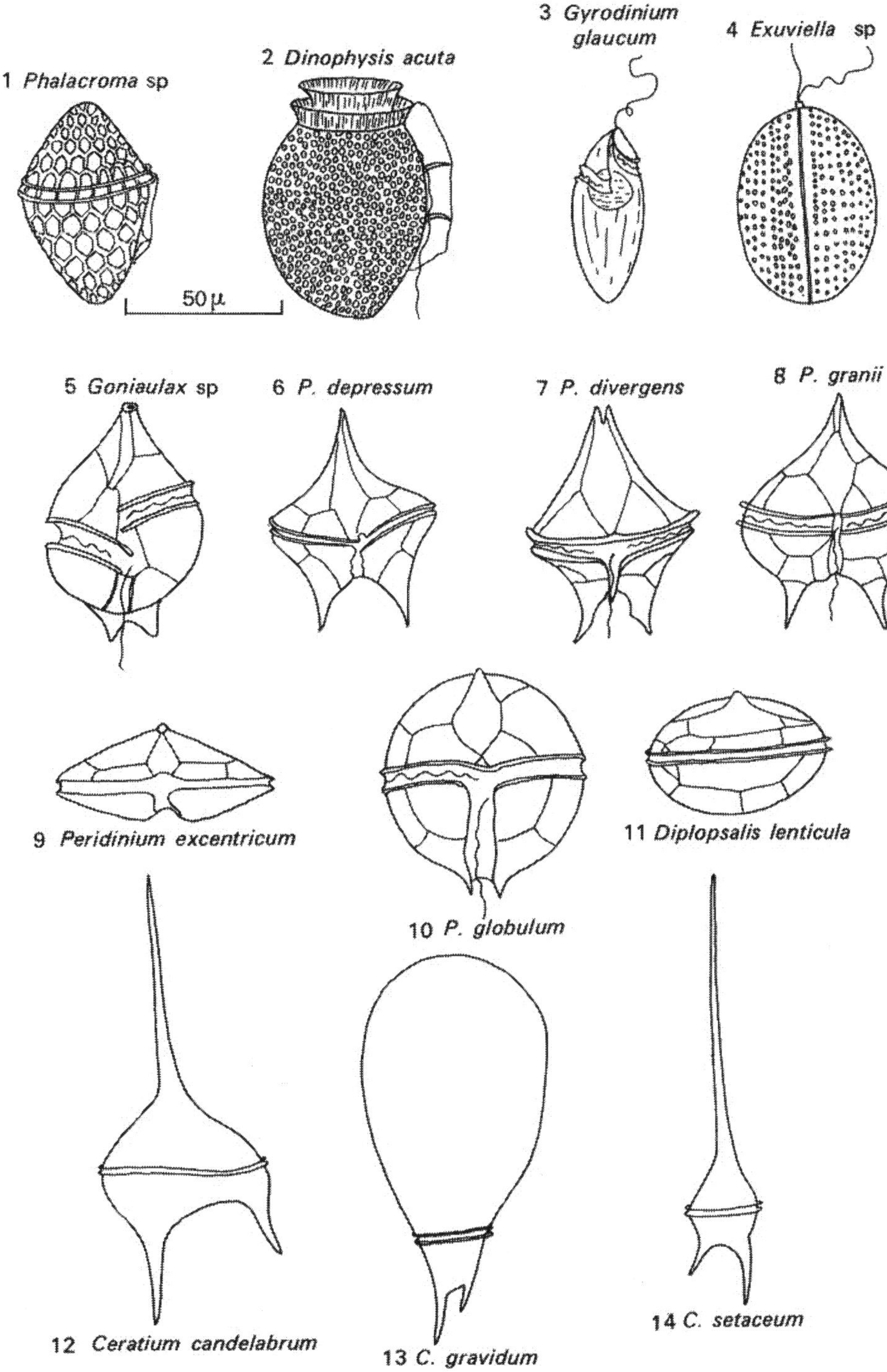

Plate 10. Dinoflagellates

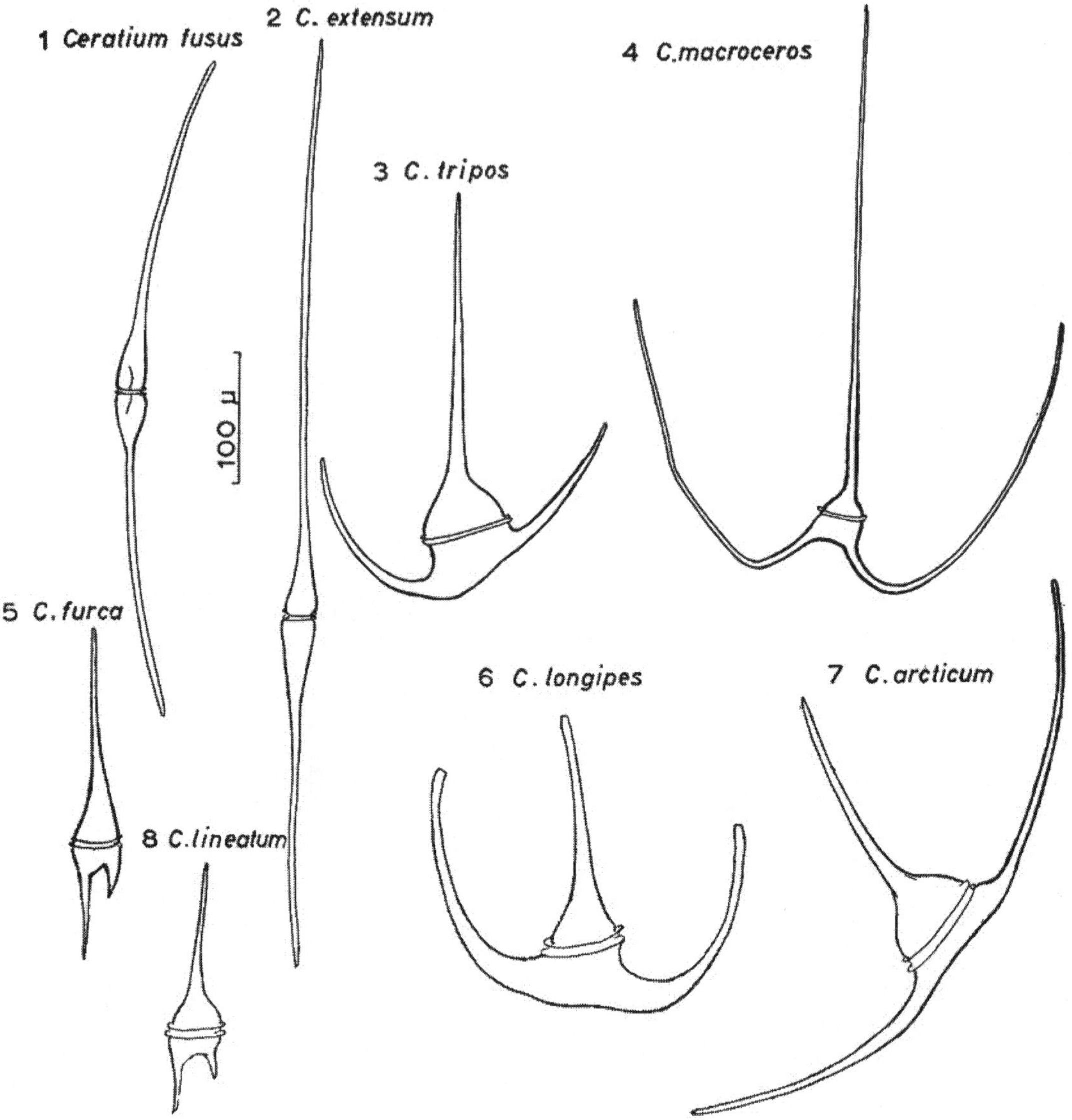

Plate 11. Dinoflagellates

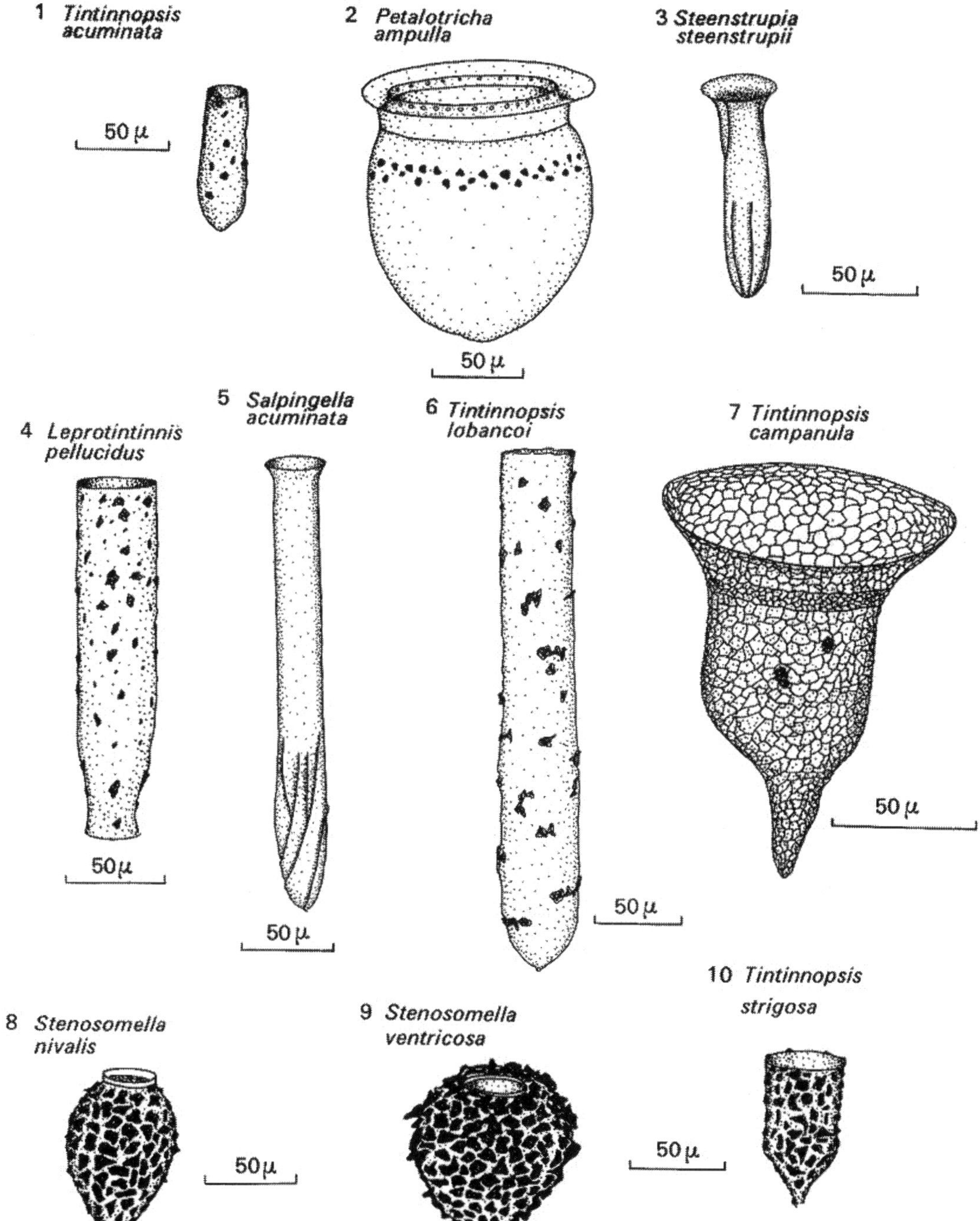

Plate 12. Tintinnids

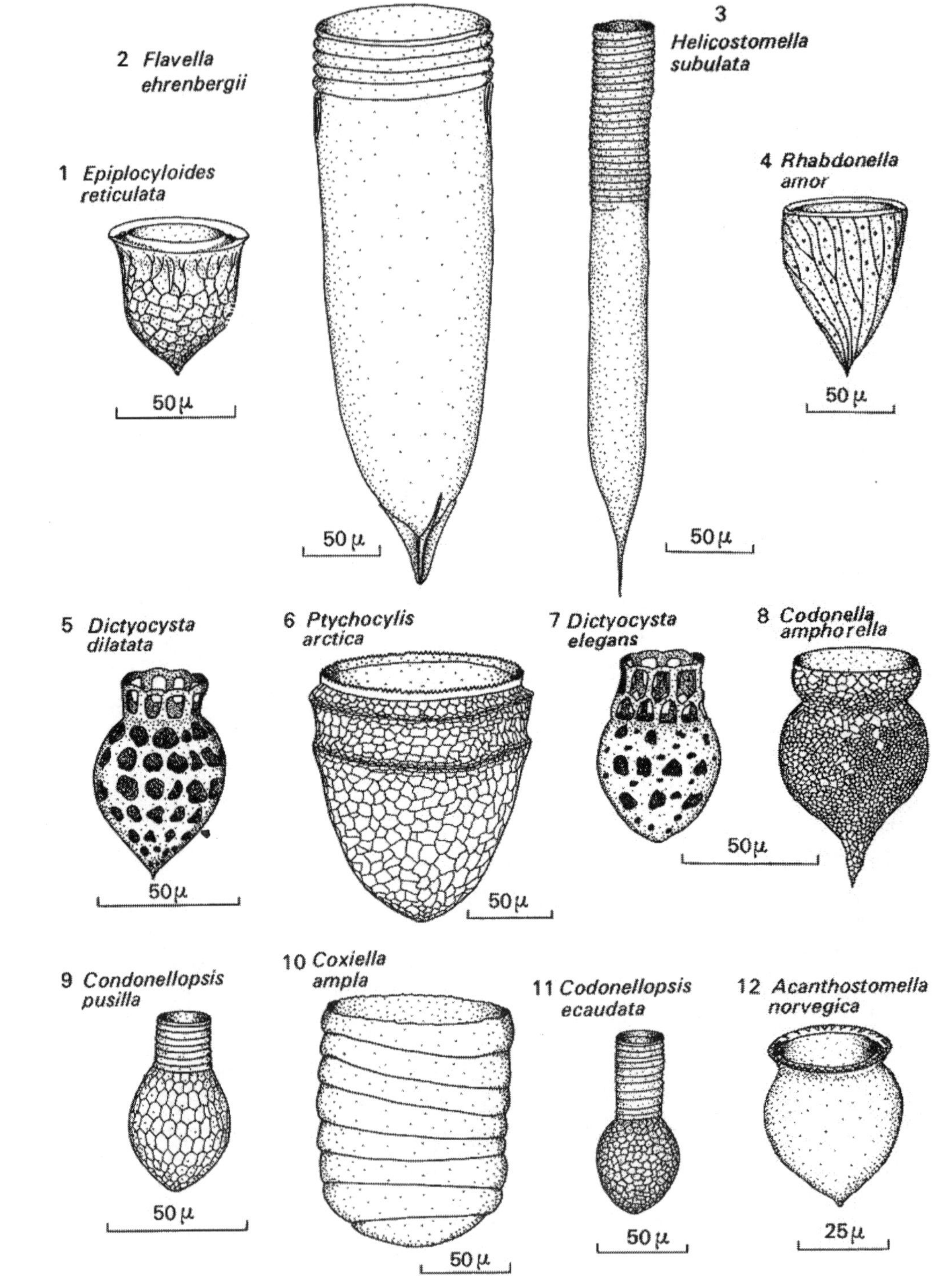

Plate 13. Tintinnids

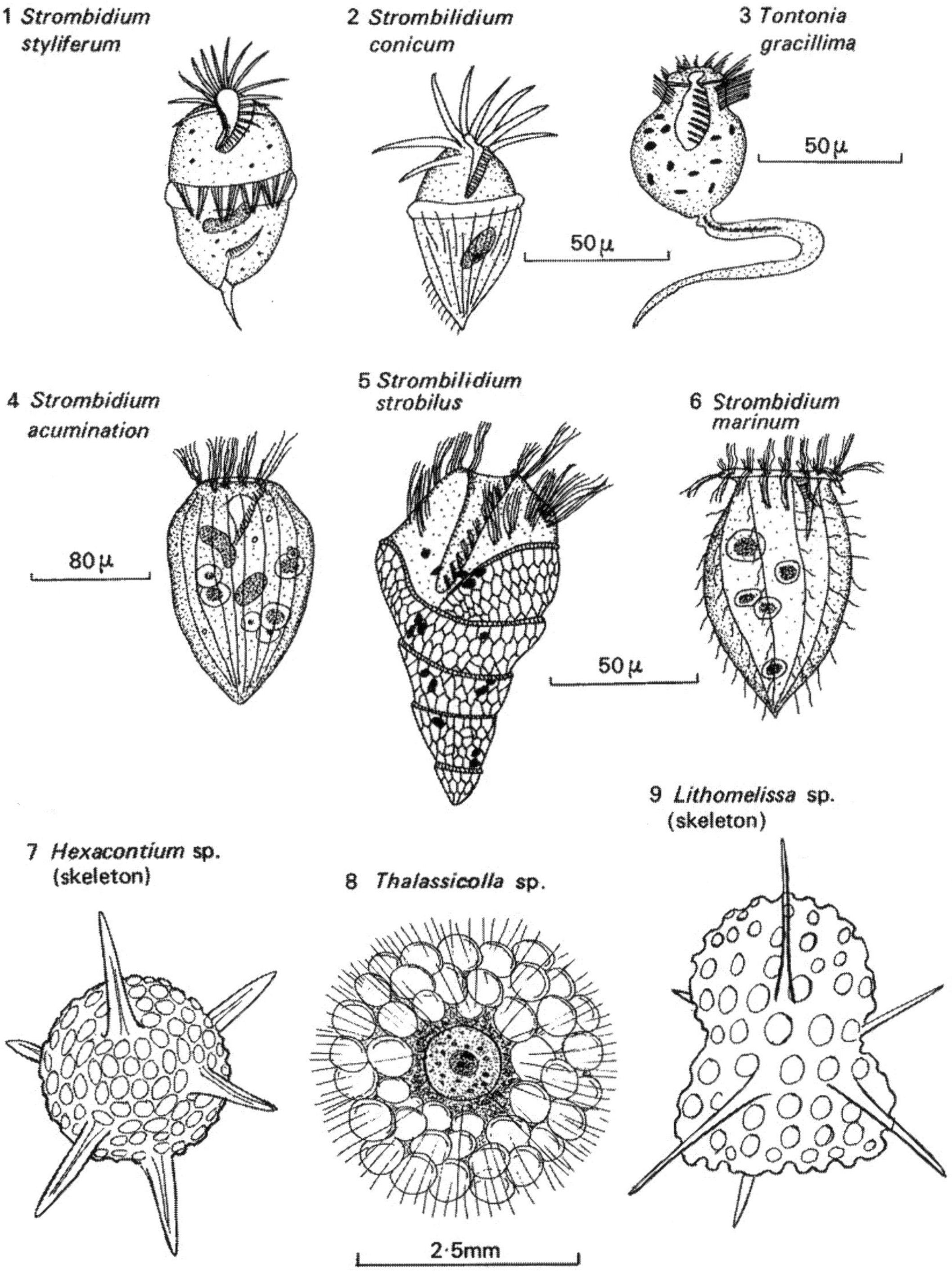

Plate 14. Oligotrichs and Radiolarians

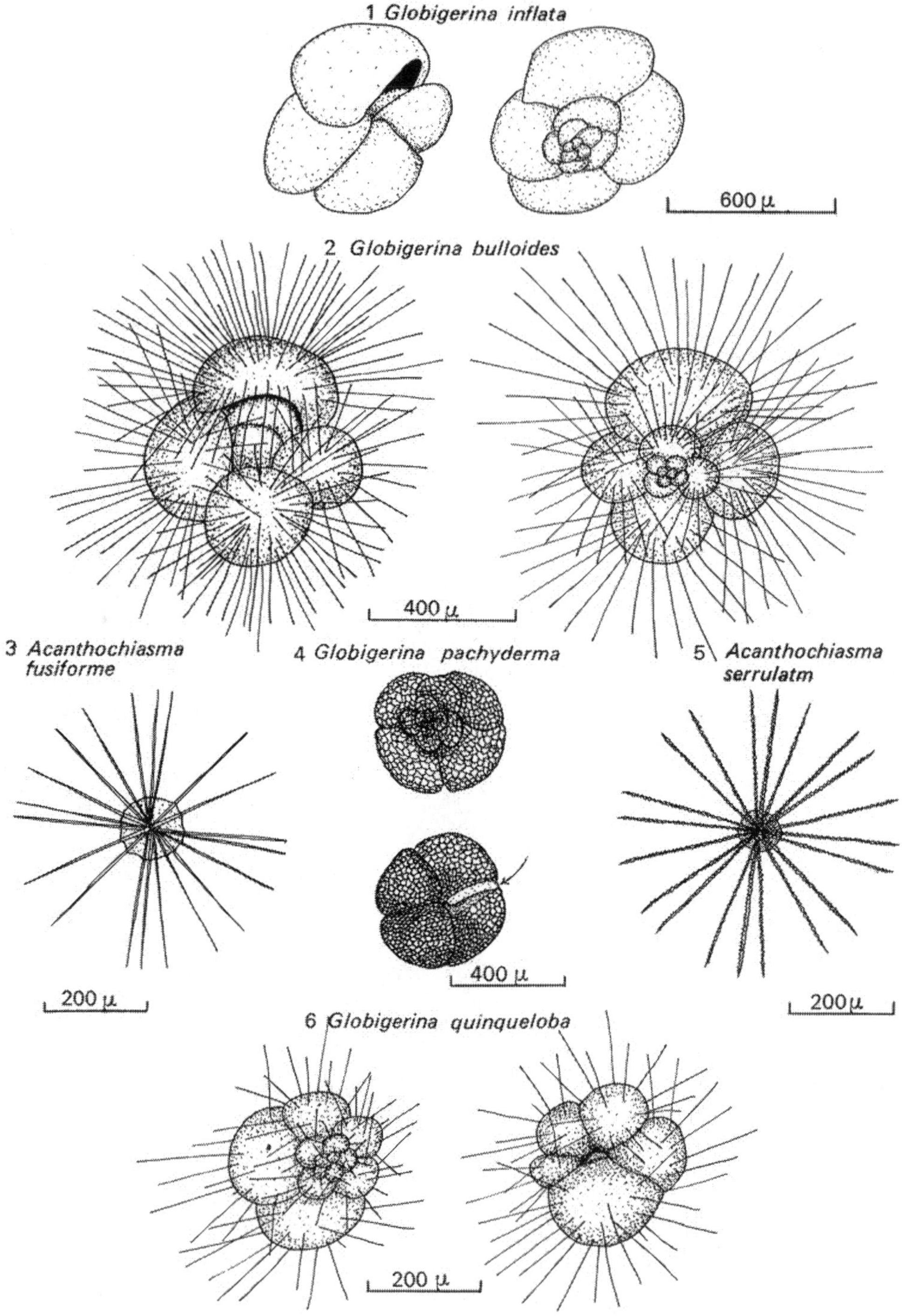

Plate 15. Foraminiferans and Acantharians

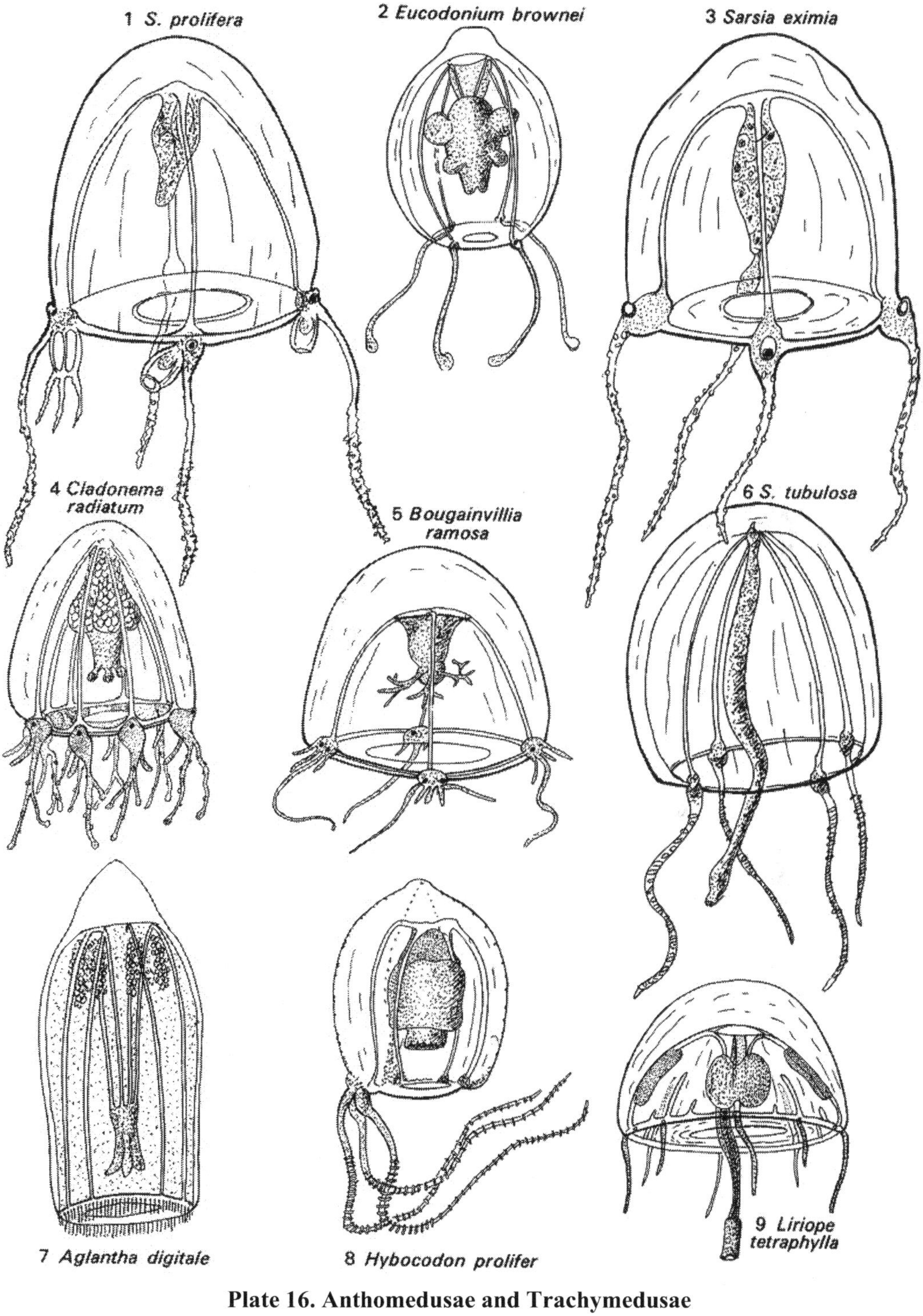

Plate 16. Anthomedusae and Trachymedusae

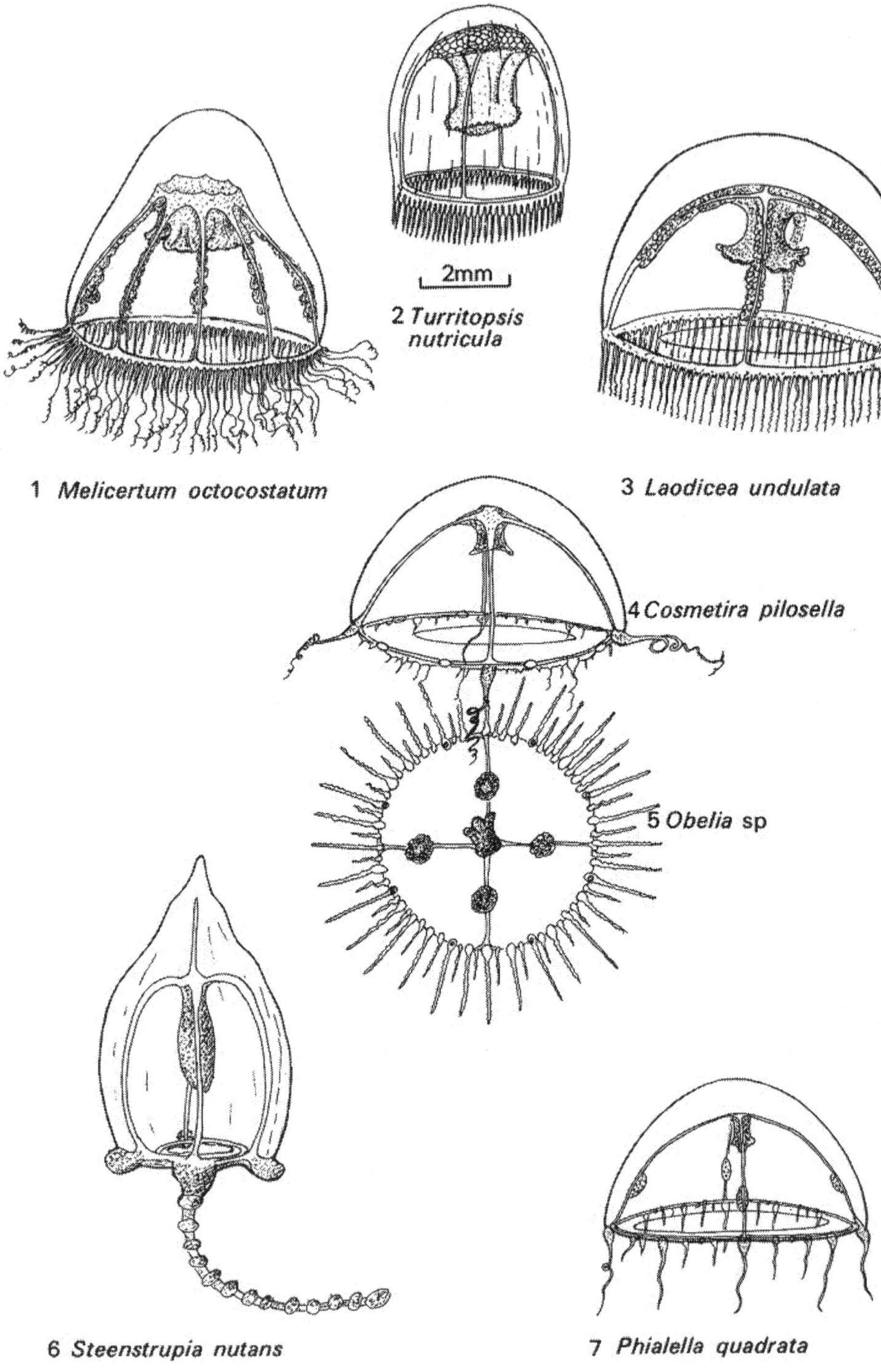

Plate 17. Anthomedusae and Leptomedusae

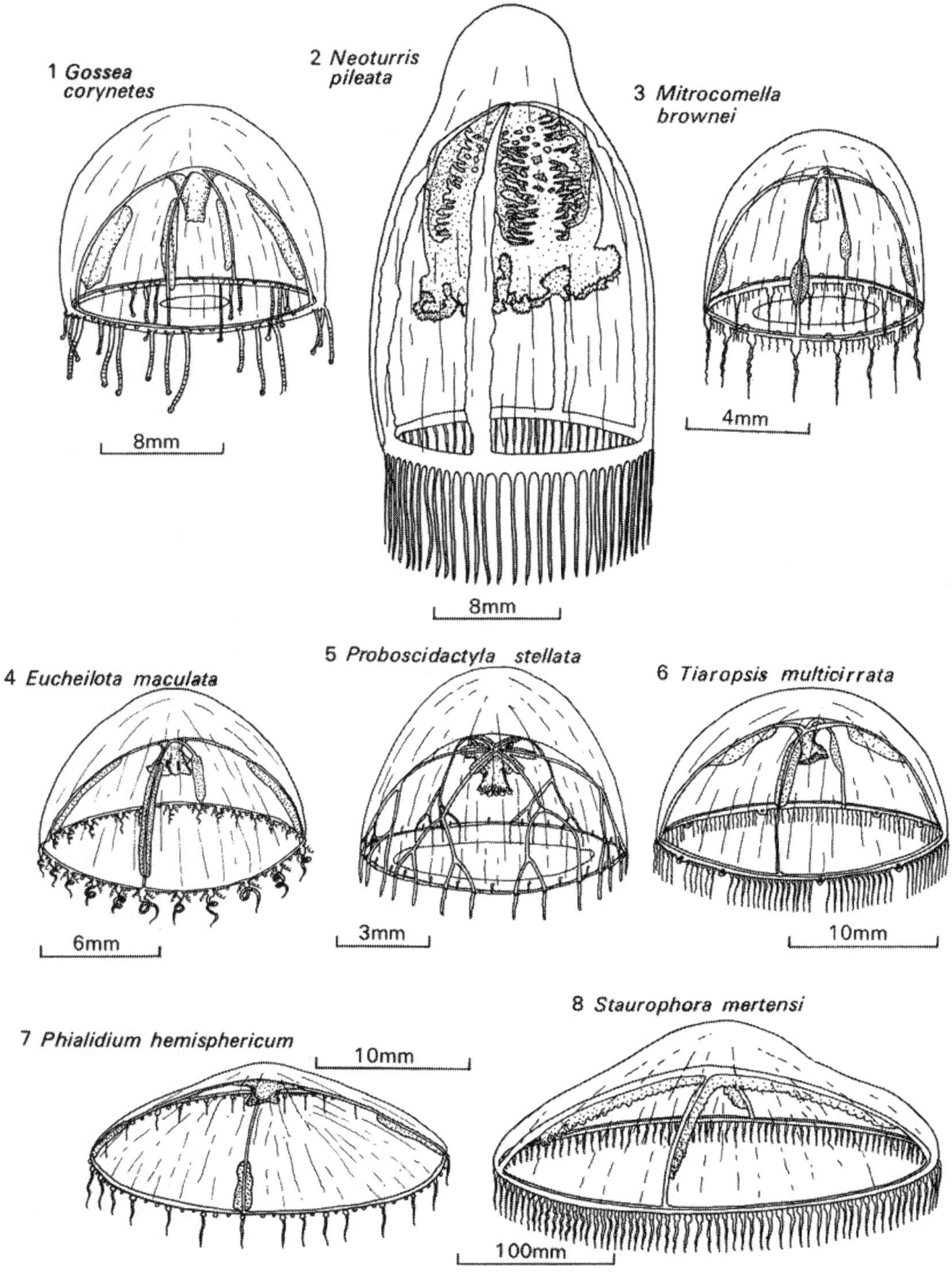

Plate 18. Leptomedusae, Anthomedusae and Limnomedusae

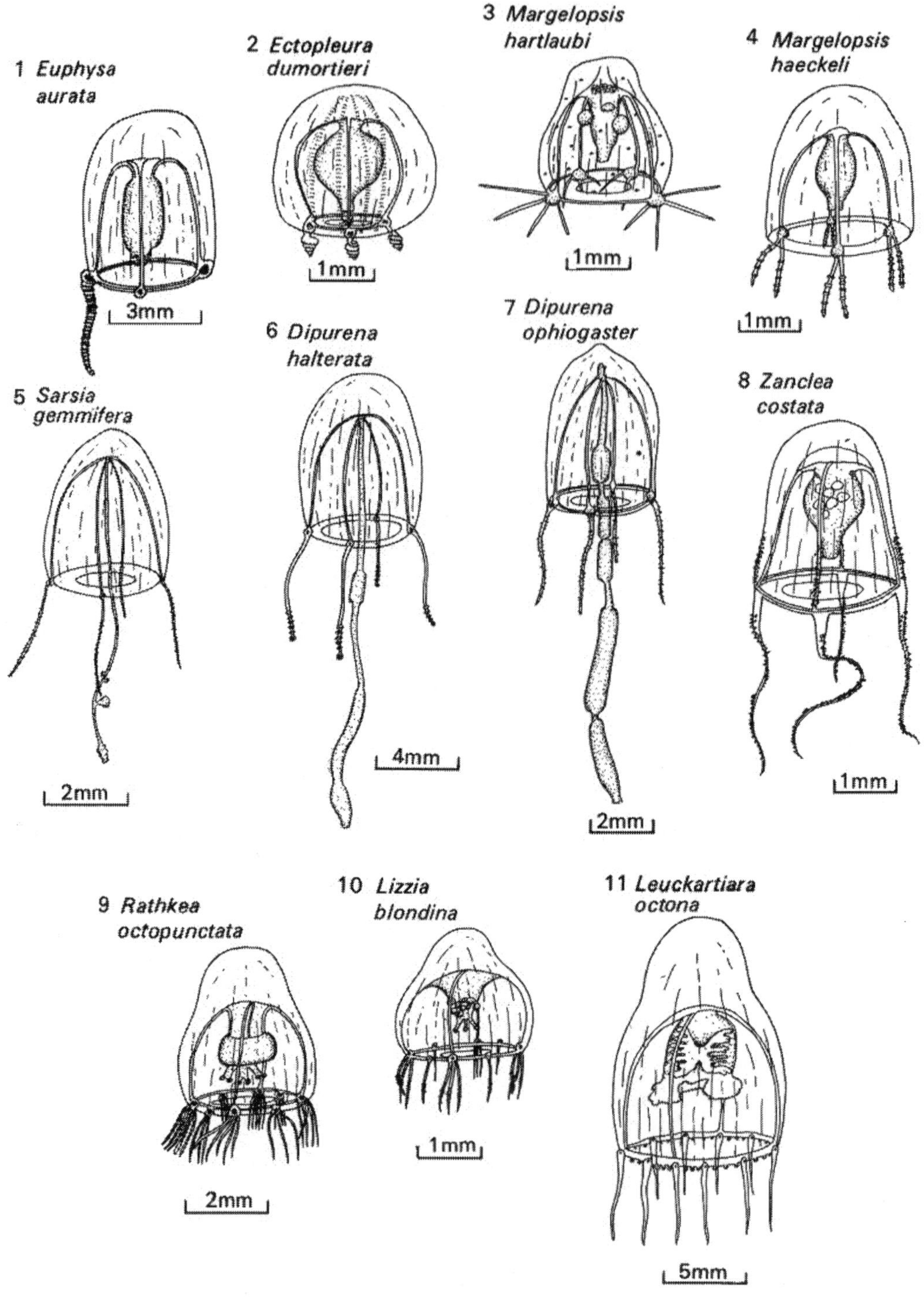

Plate 19. Anthomedusae

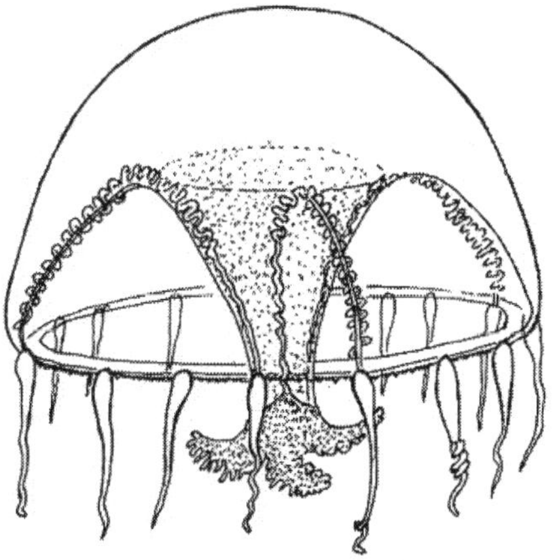

1 *Tima bairdi*

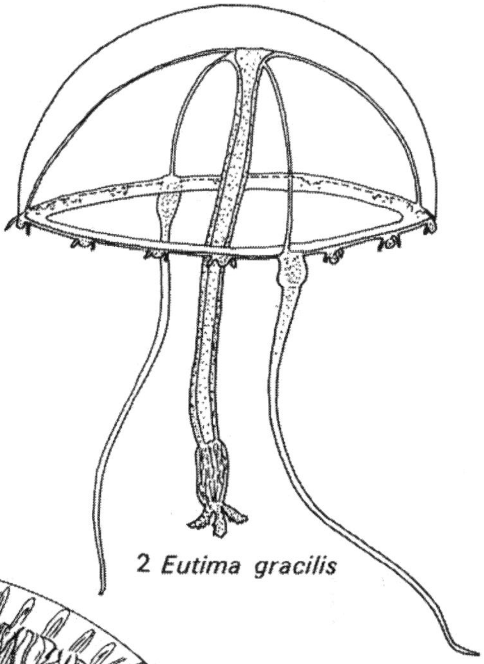

2 *Eutima gracilis*

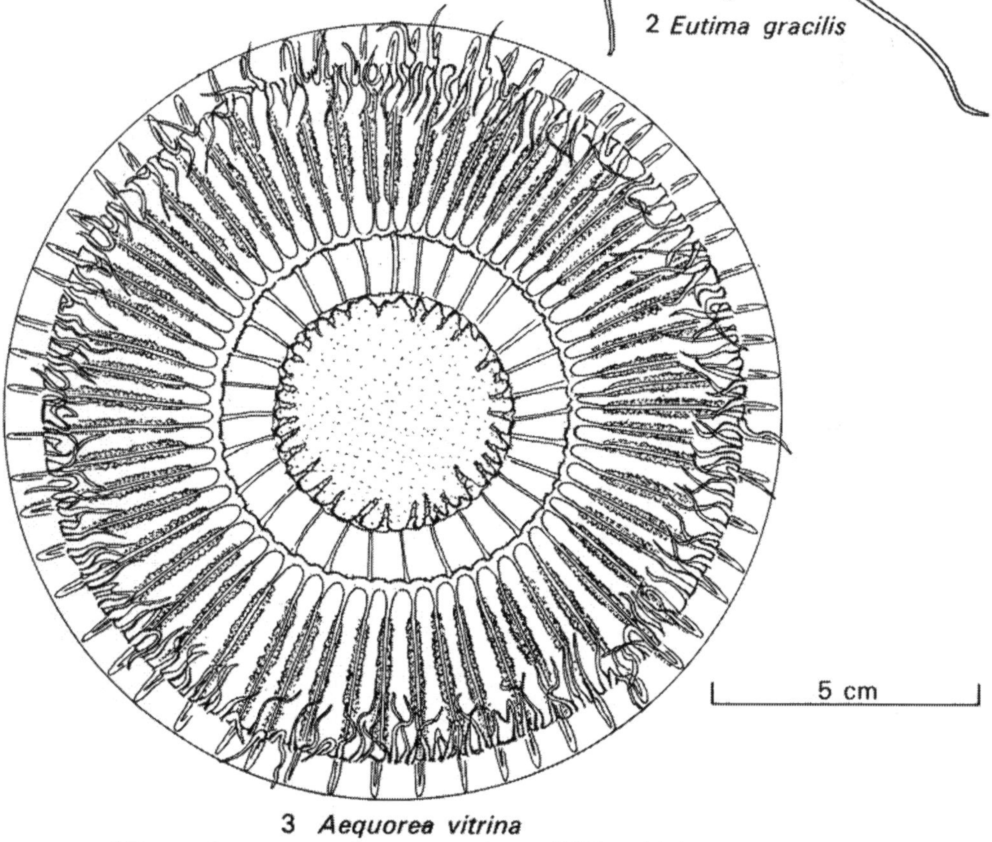

3 *Aequorea vitrina*
(drawn from a specimen caught at Whitstable)

Plate 20. Leptomedusae

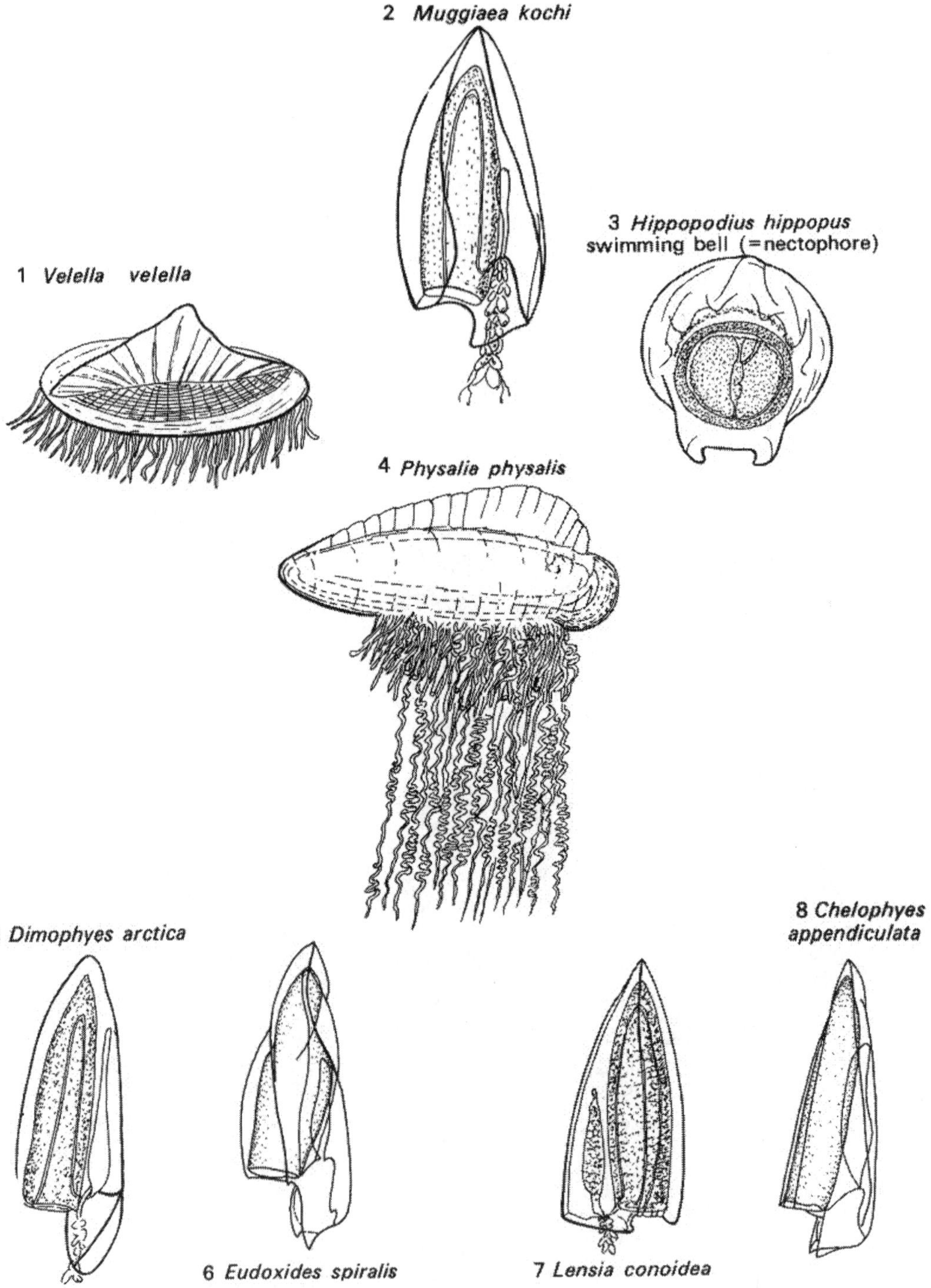

Plate 21. Siphonophores

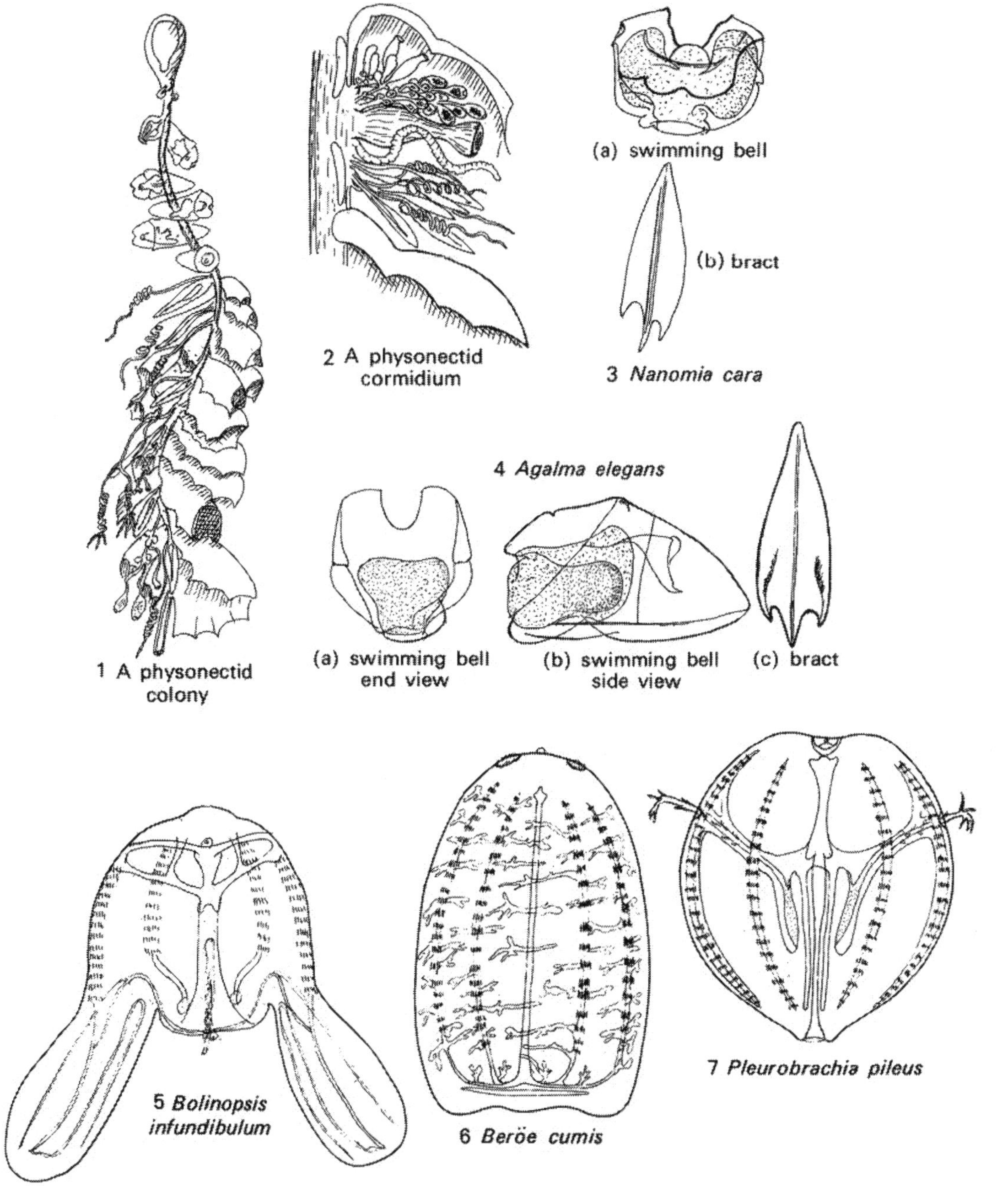

Plate 22. Siphonophores and Ctenophores

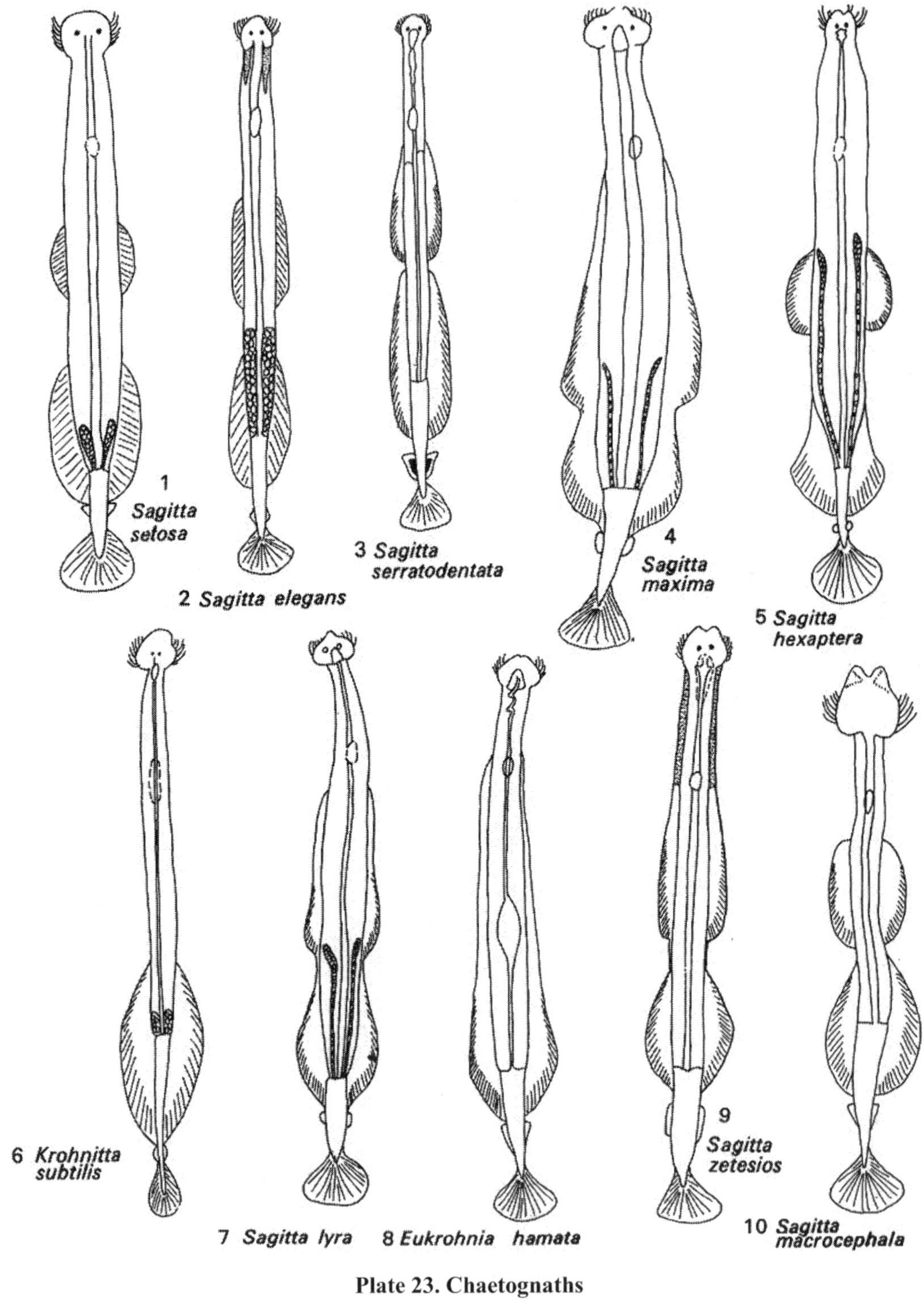

Plate 23. Chaetognaths

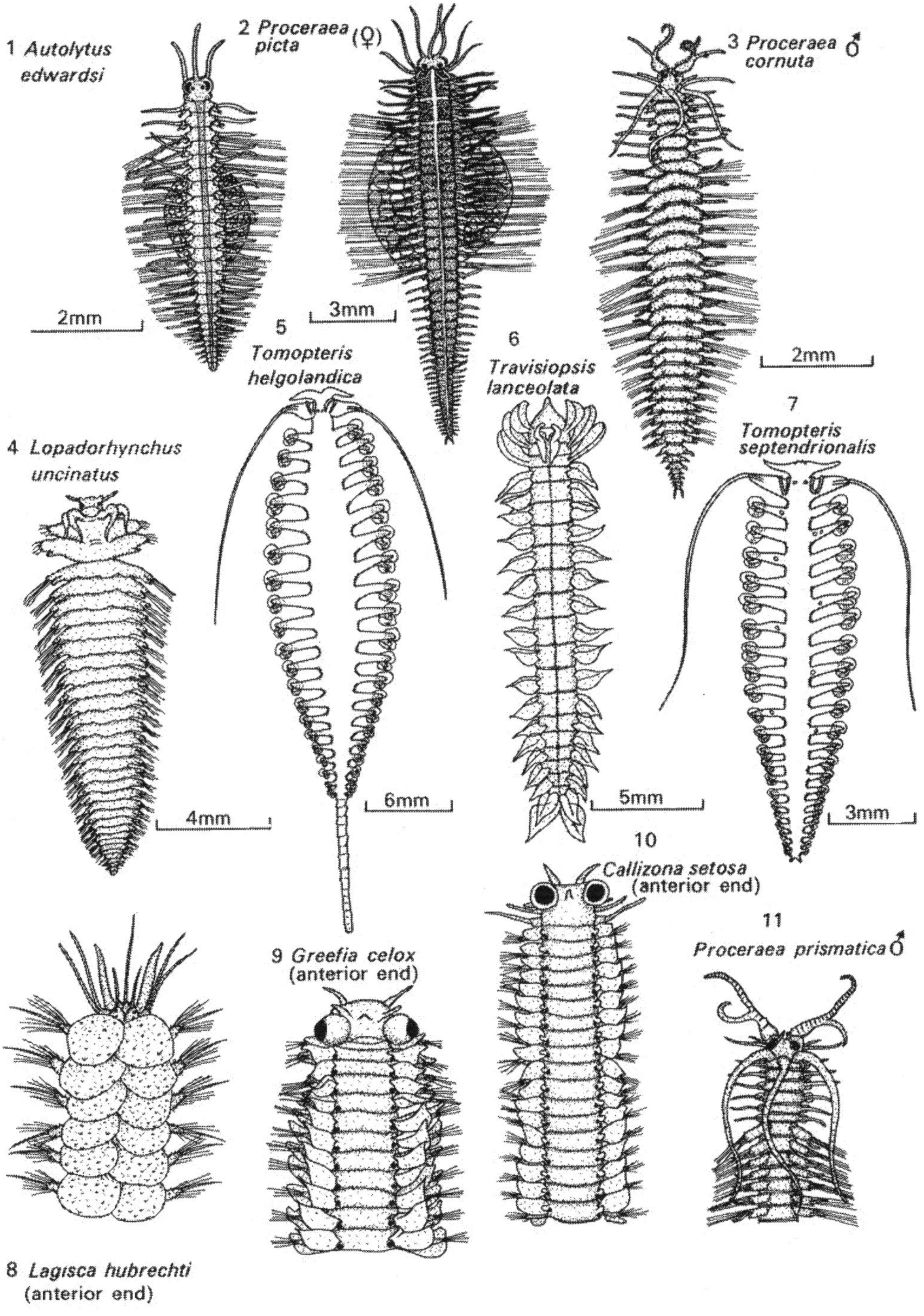

Plate 24. Polychaetes

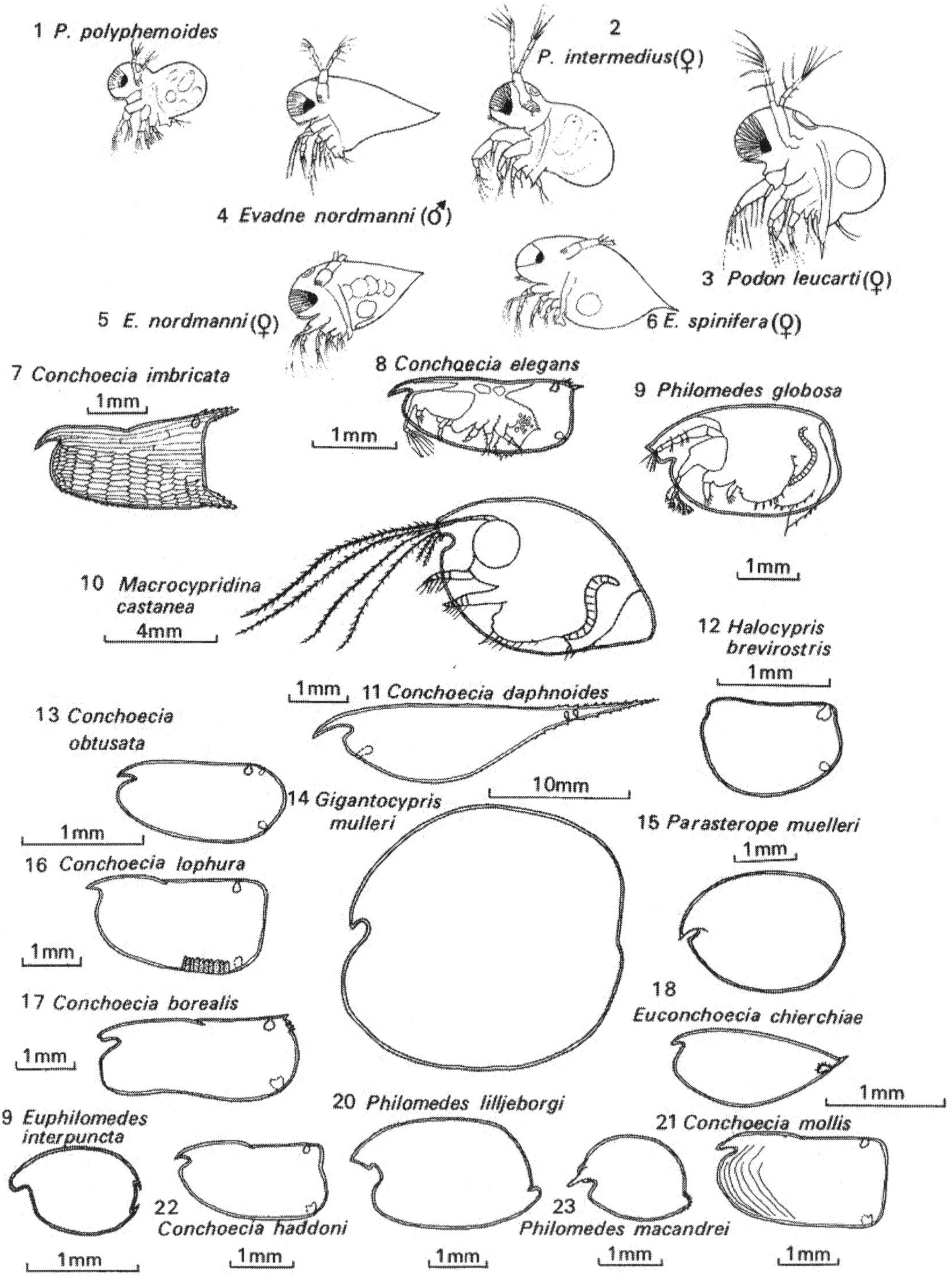

Plate 25. Cladocera and Ostracods

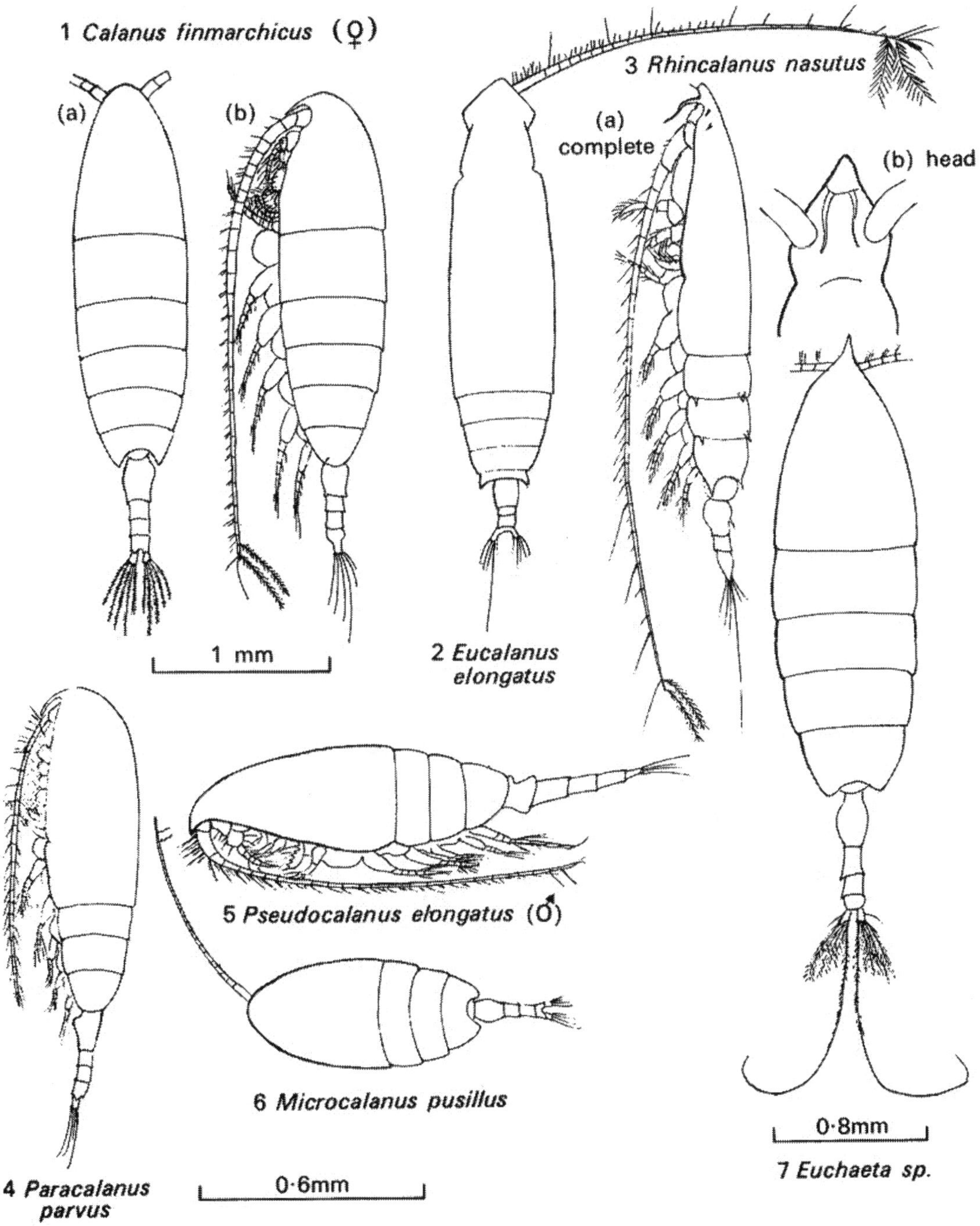

Plate 26. Calanoid Copepods

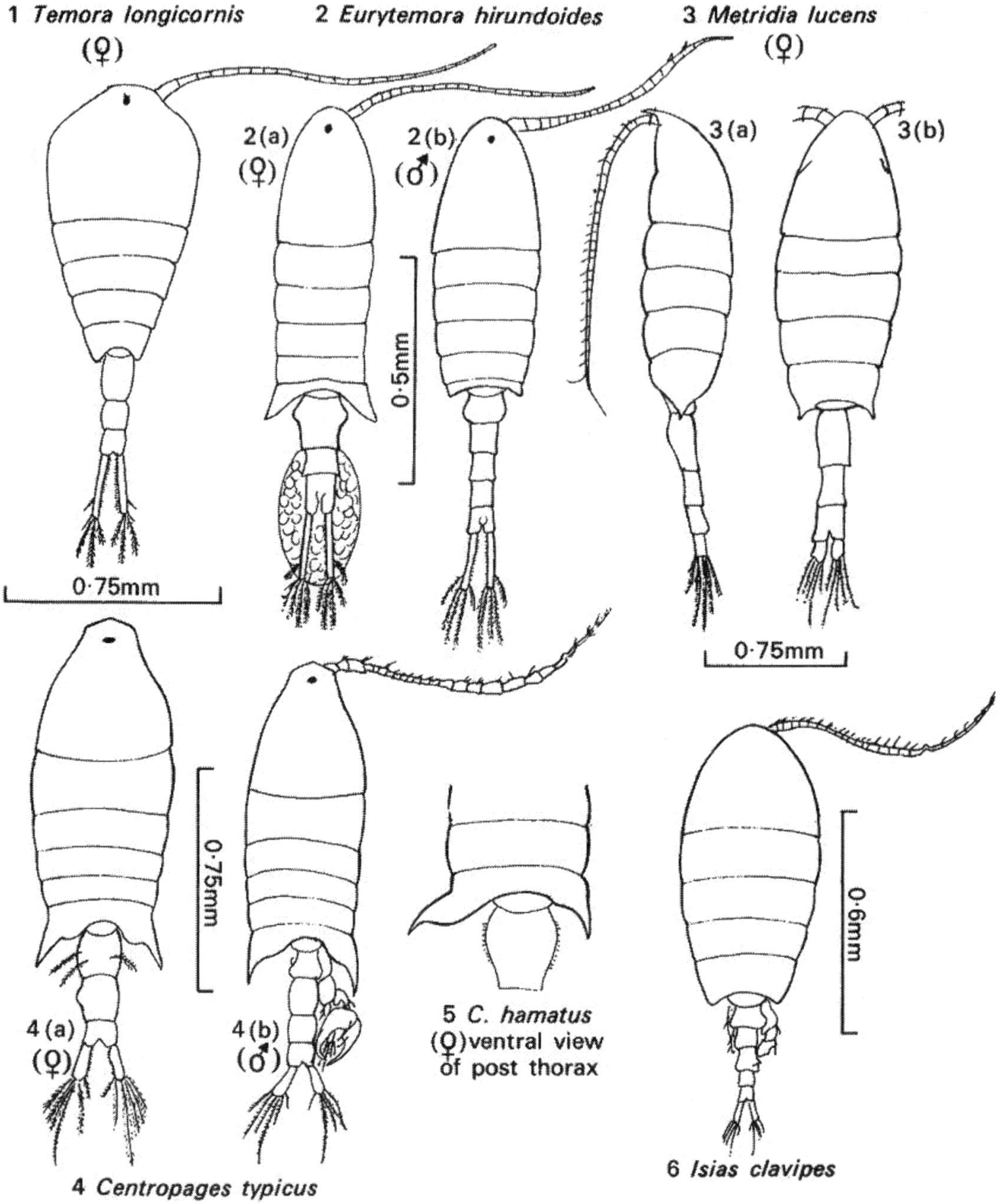

Plate 27. Calanoid Copepods

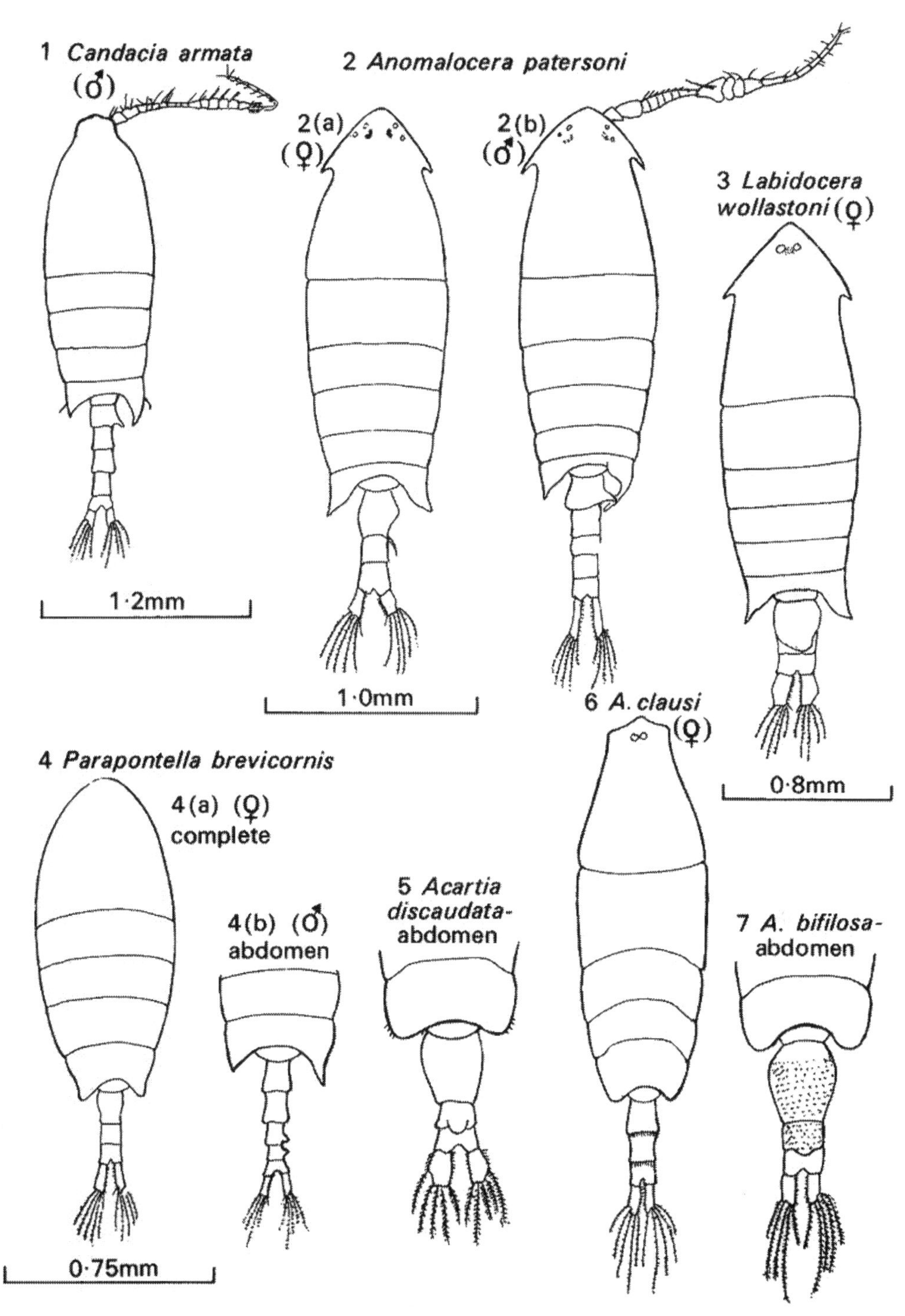

Plate 28. Calanoid Copepods

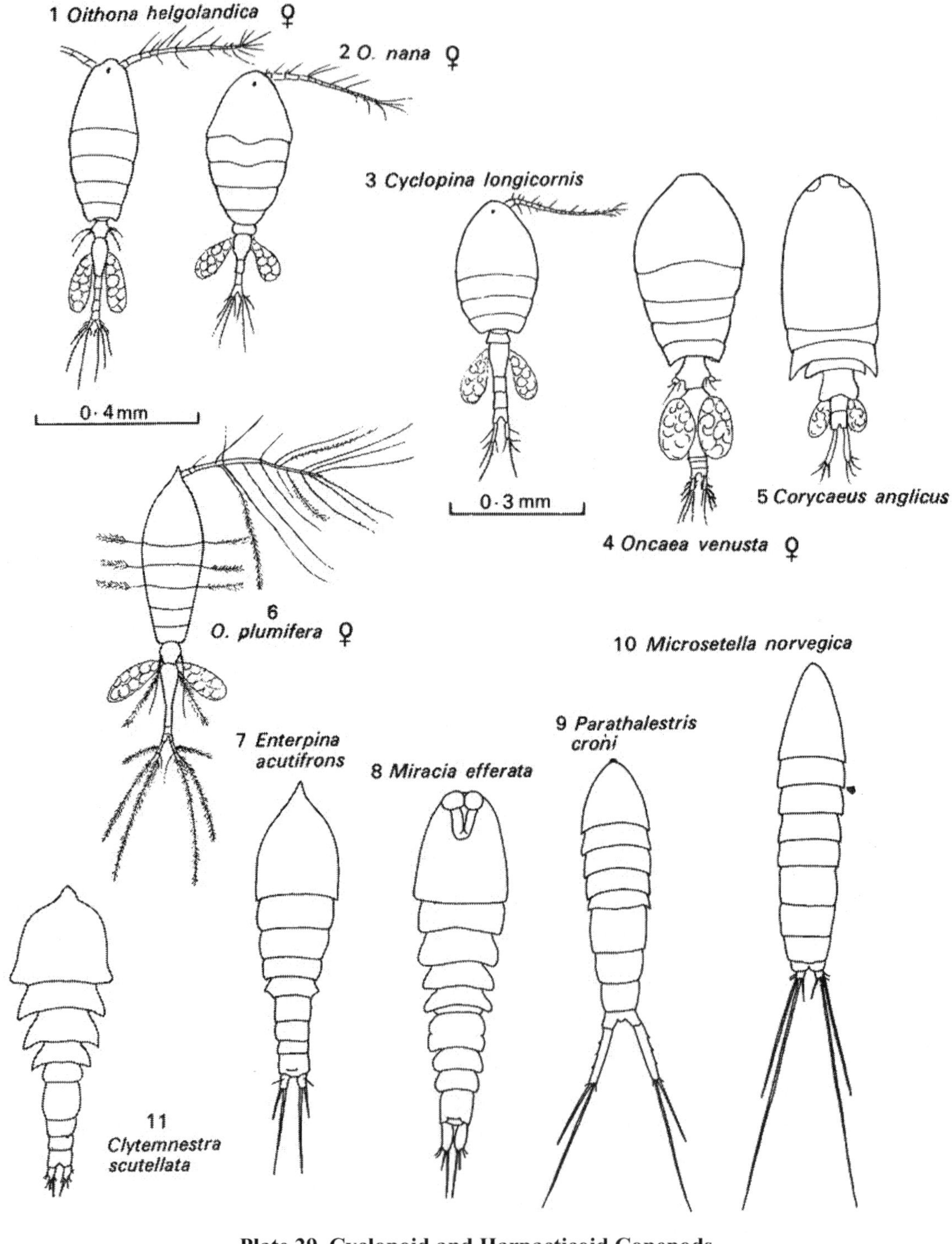

Plate 29. Cyclopoid and Harpacticoid Copepods

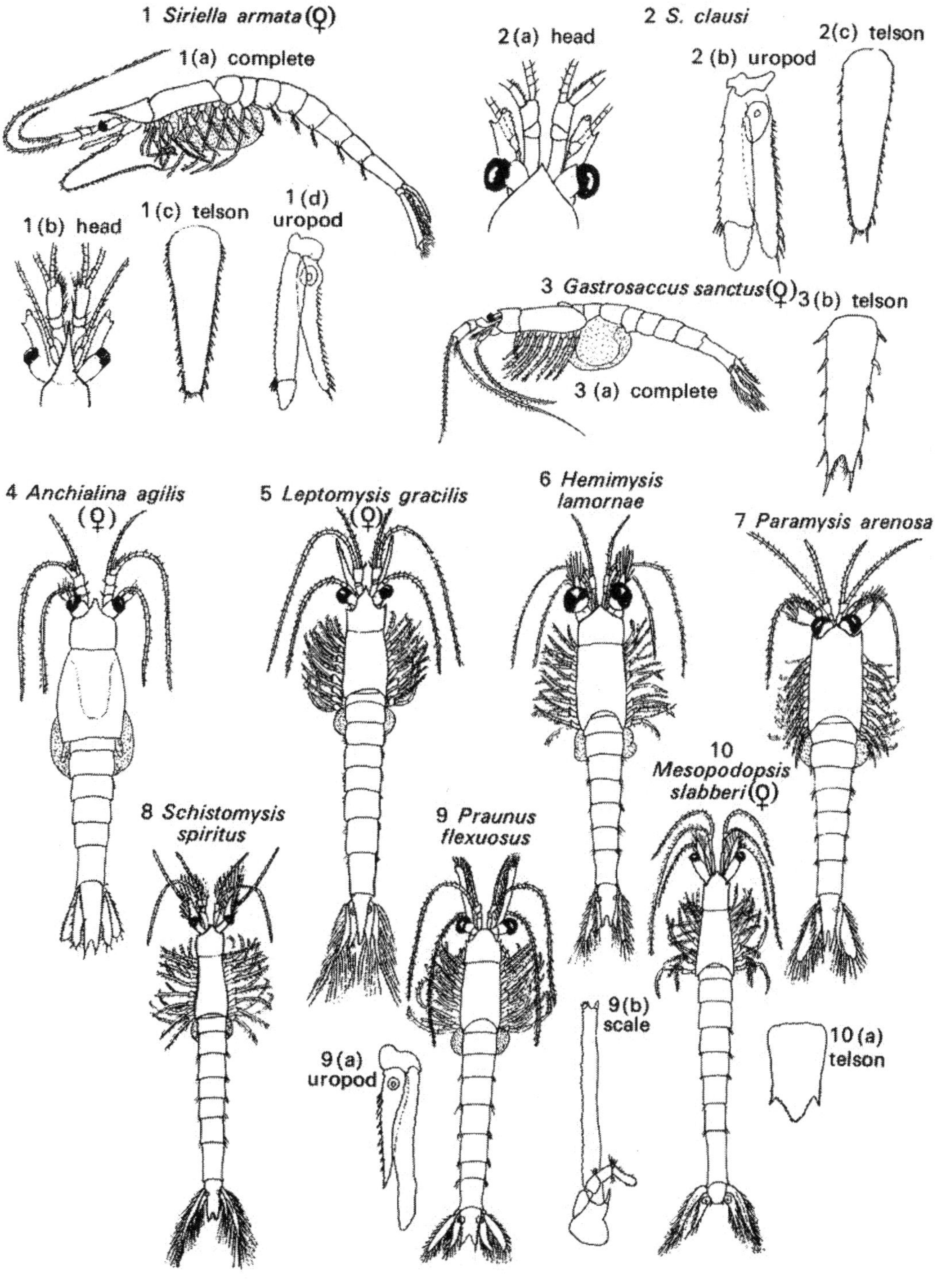

Plate 30. Mysids

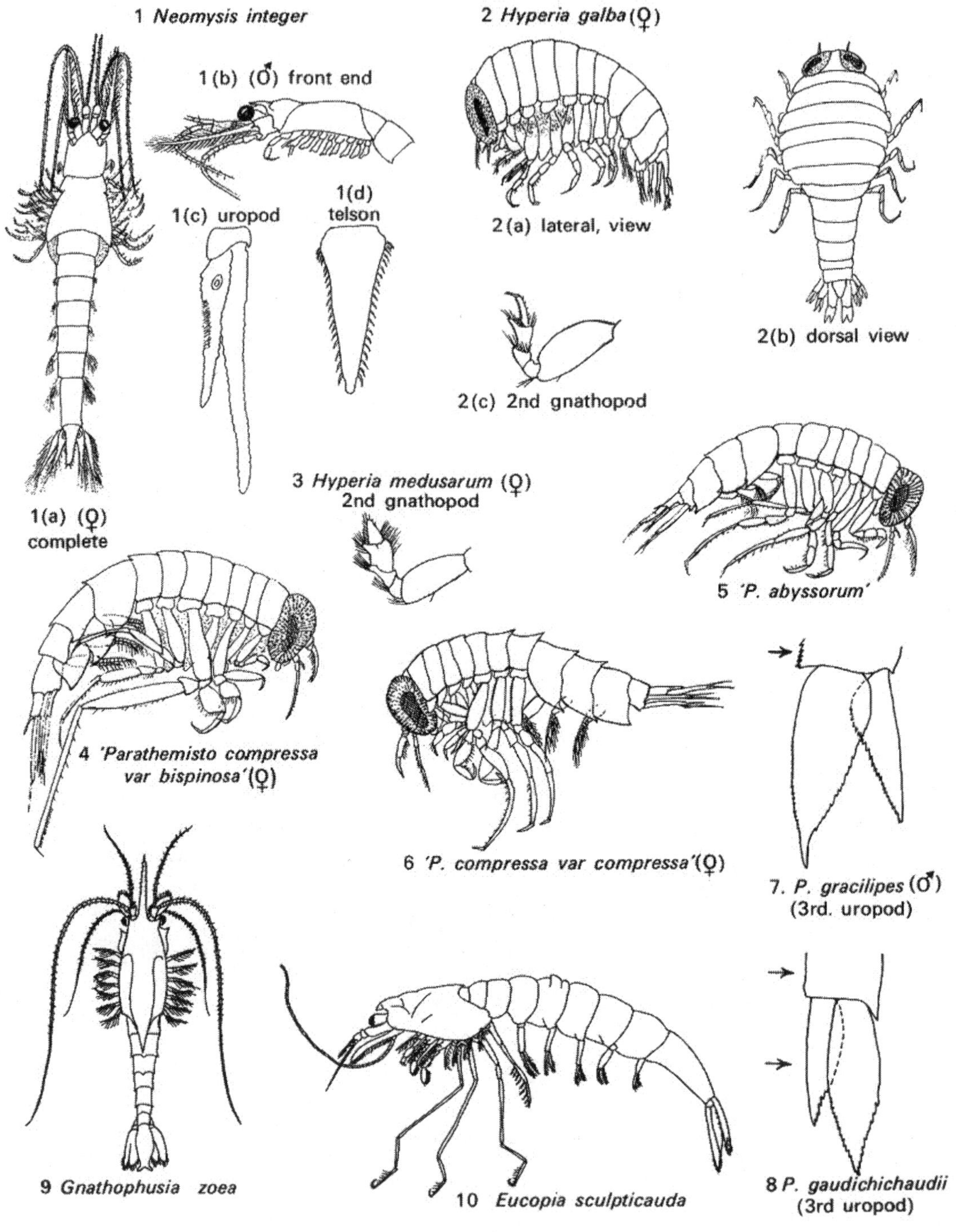

Plate 31. Mysids and Hyperiids

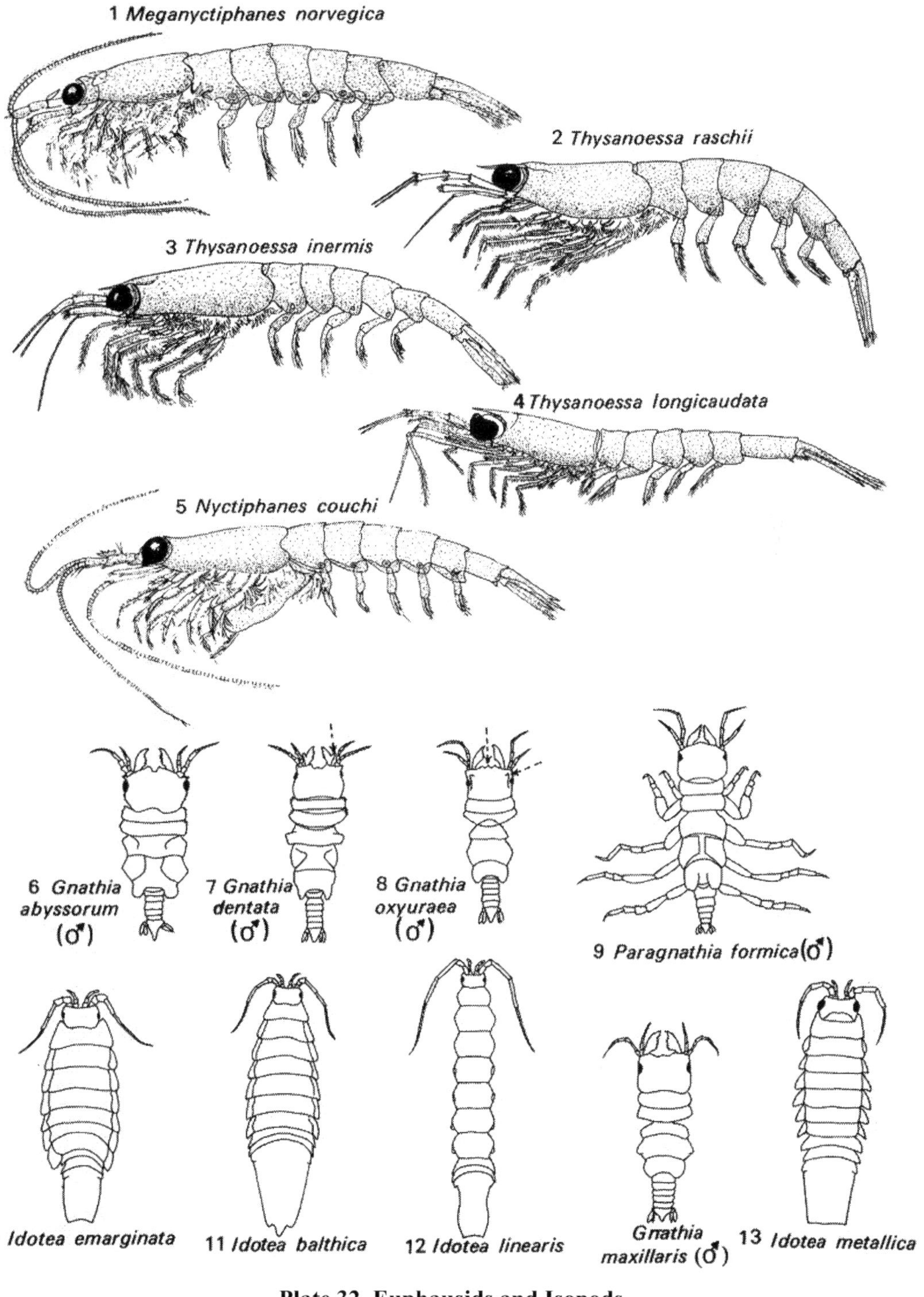

Plate 32. Euphausids and Isopods

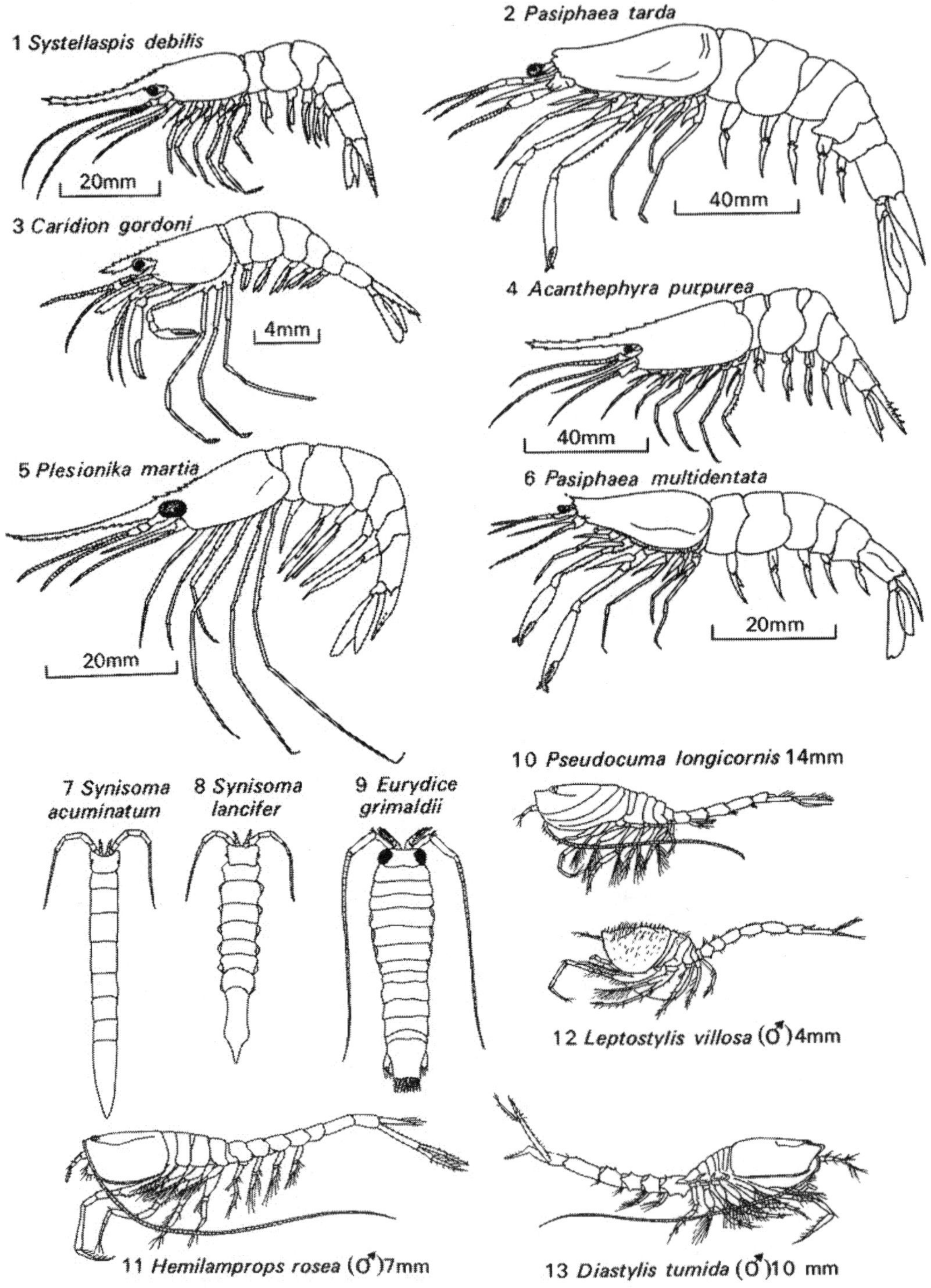

Plate 33. Decapods, Isopods and Cumaceans

(a) aggregate form

(b) solitary form

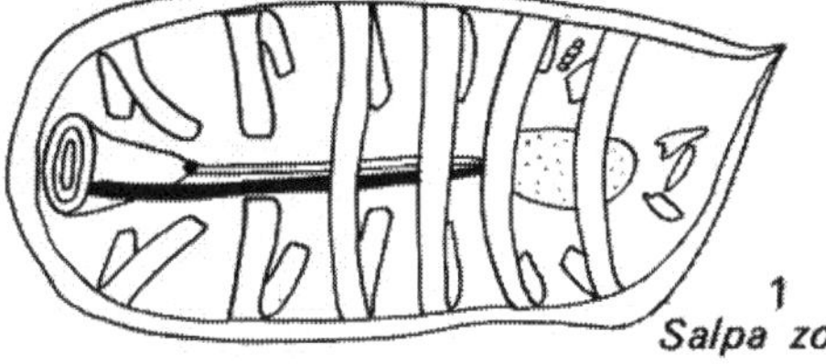

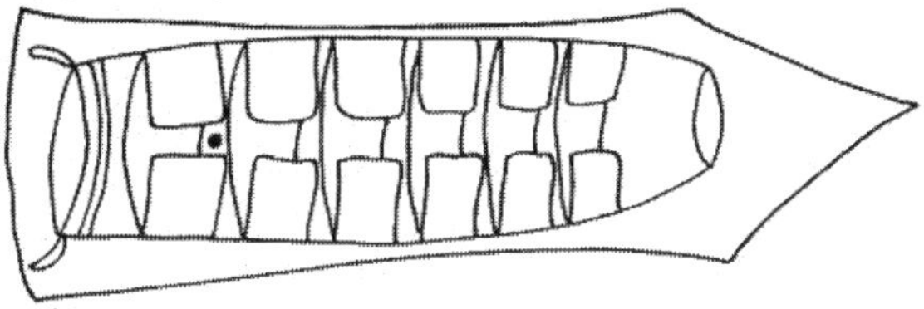

1
Salpa zonaria

2 *Salpa fusiformis*

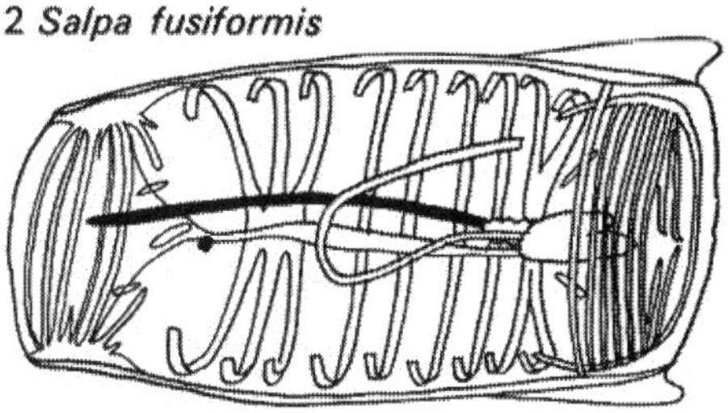

(a) oozooid from the ventral surface

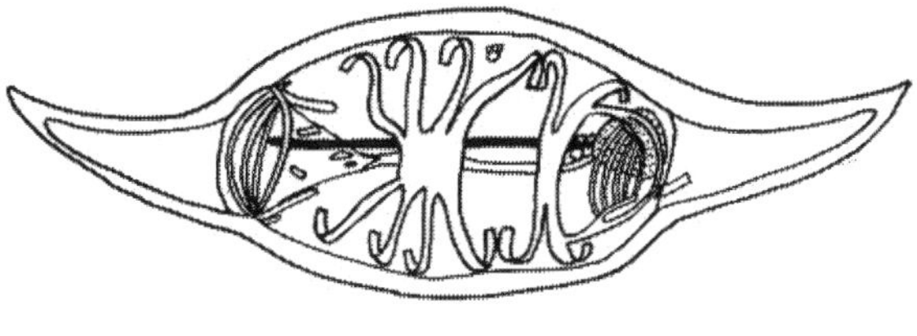

(b) blastozooid

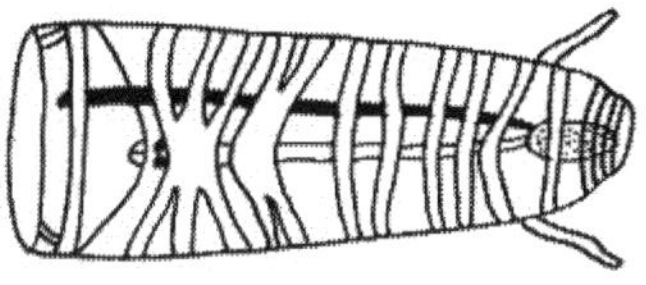

(a) solitary form

3 *Salpa asymmetrica*

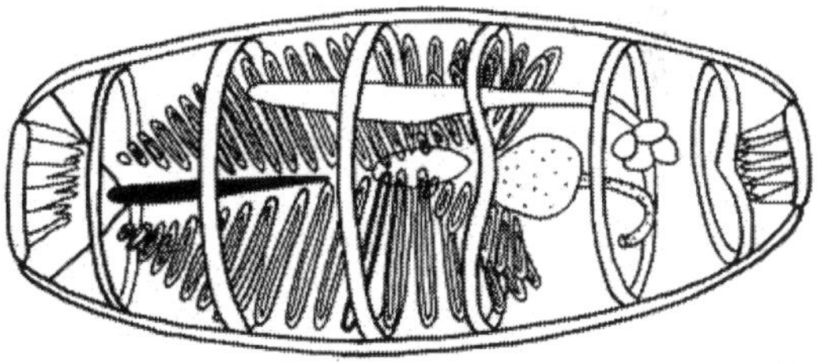

4 *D. nationalis* gonozooid from the dorsal surface

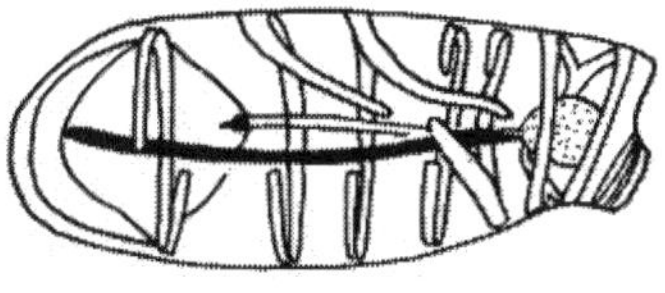

(b) aggregate form

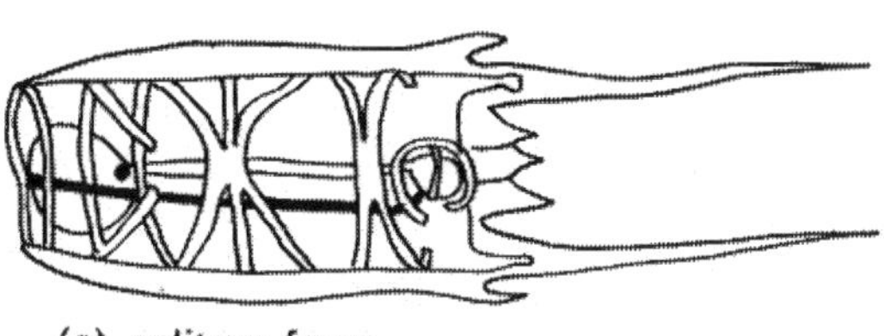

(a) solitary form

5 *Salpa democratica*

(b) aggregate form

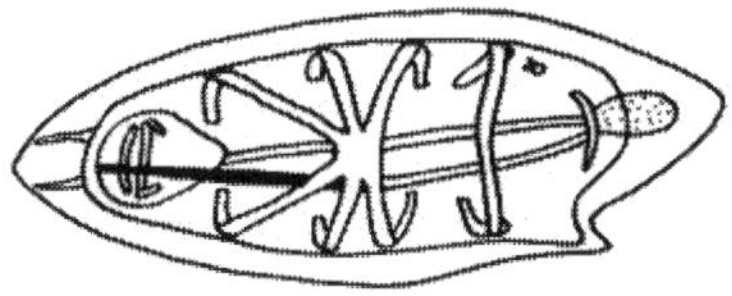

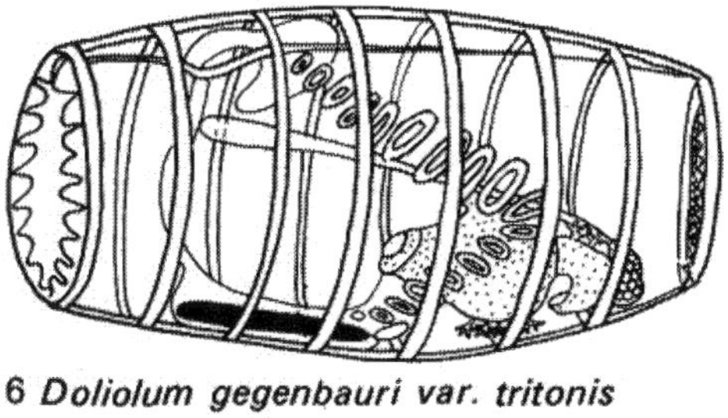

6 *Doliolum gegenbauri var. tritonis*
gonozooid from the left

Plate 34. Salps and Doliolids

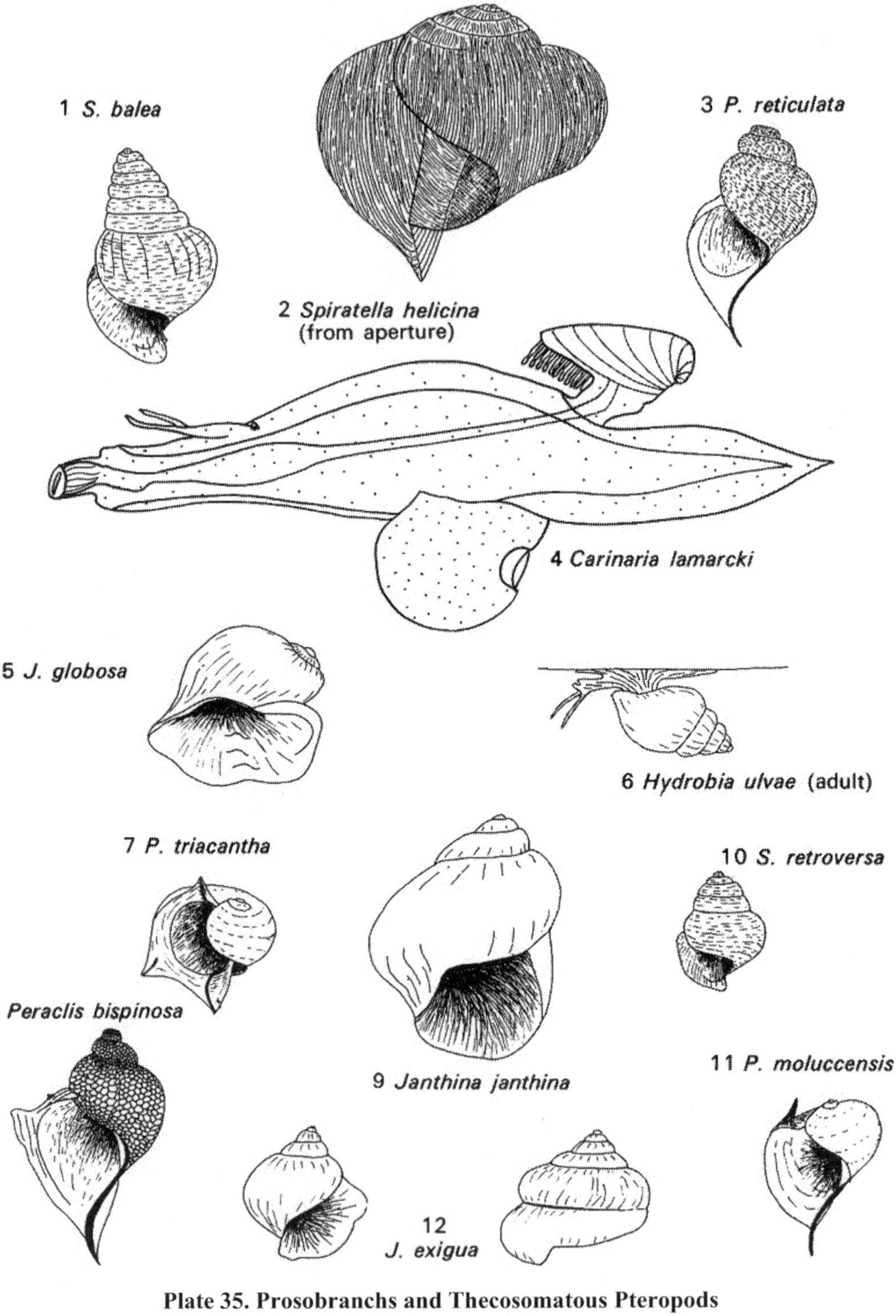

Plate 35. Prosobranchs and Thecosomatous Pteropods

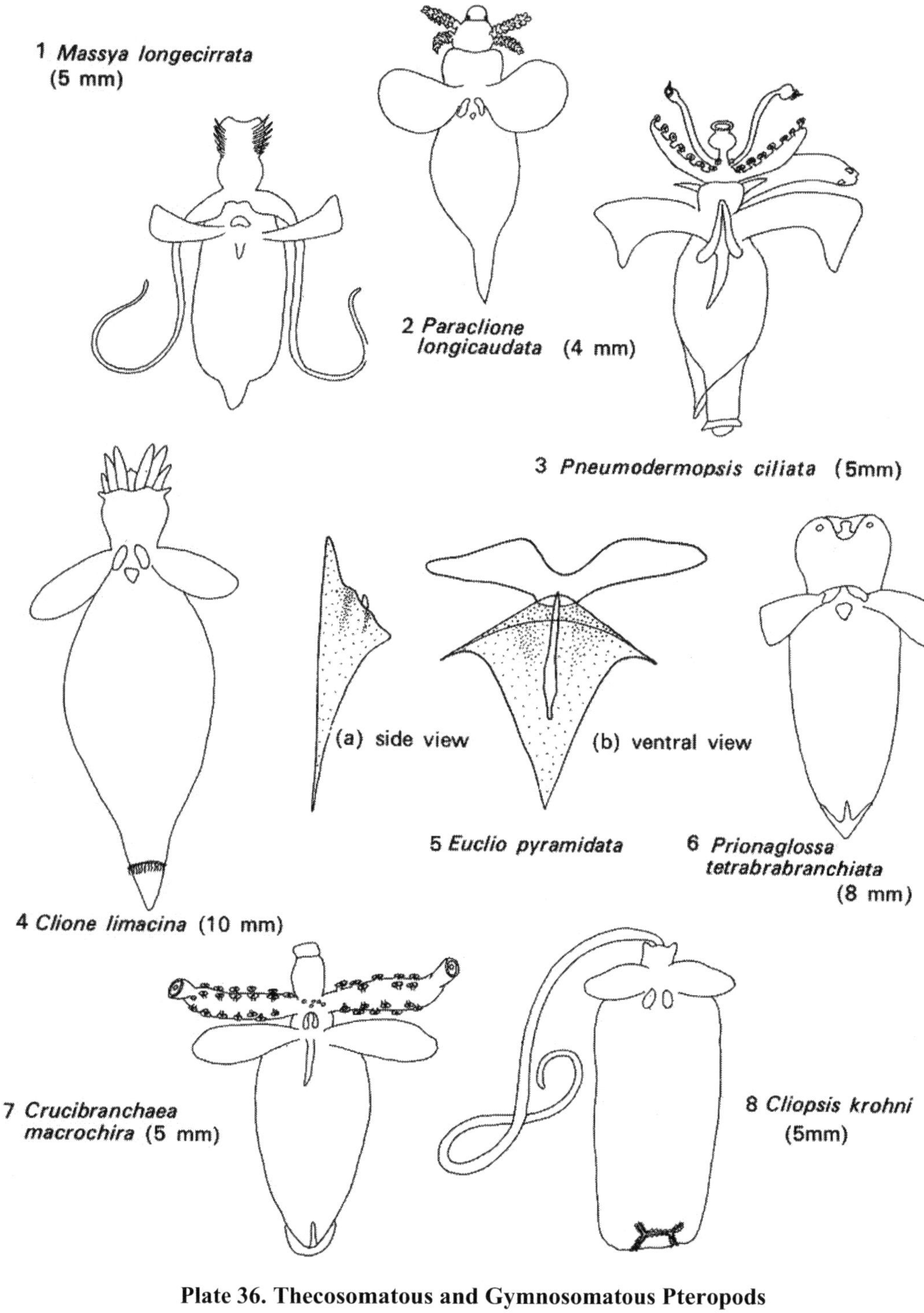

Plate 36. Thecosomatous and Gymnosomatous Pteropods

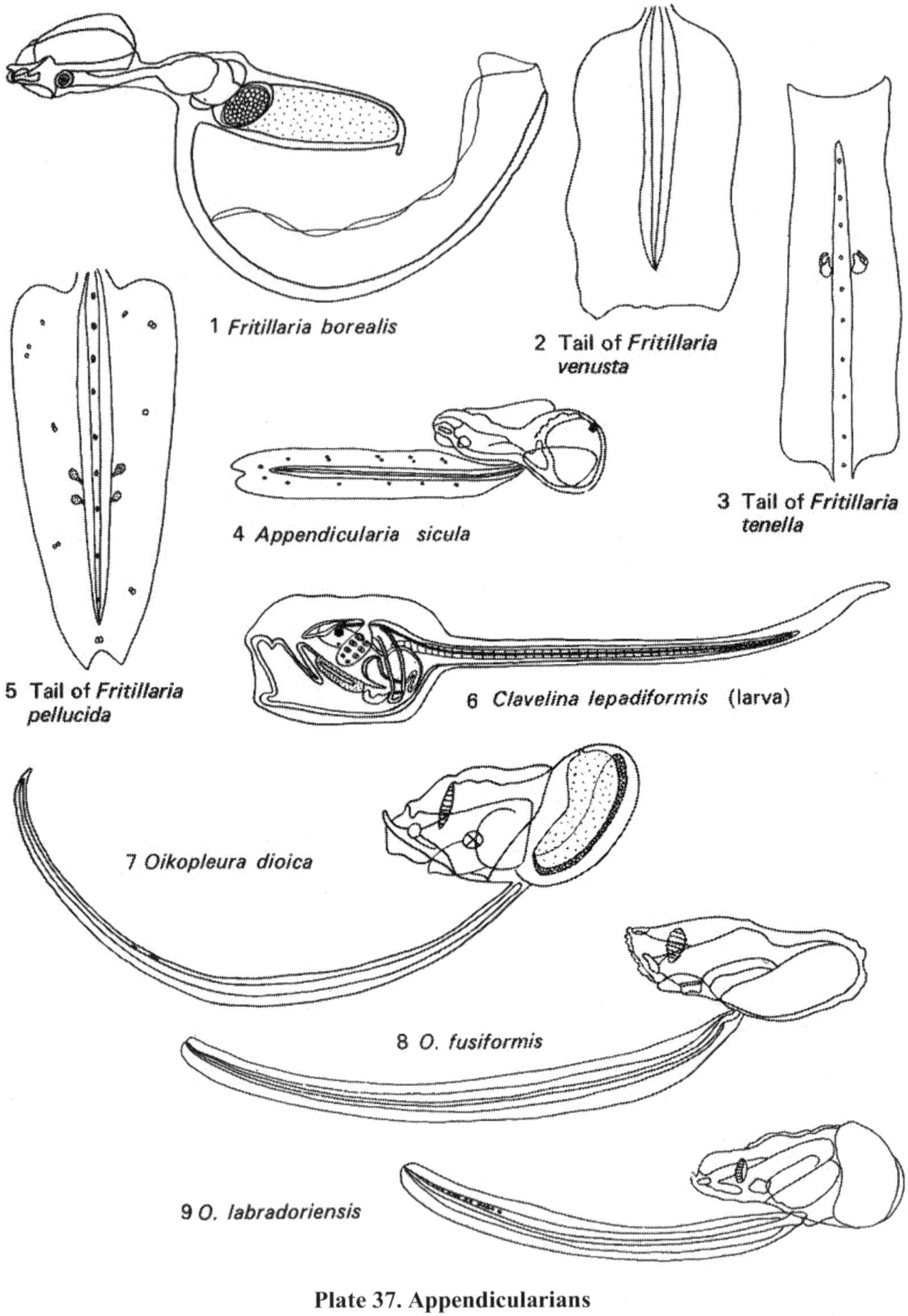

Plate 37. Appendicularians

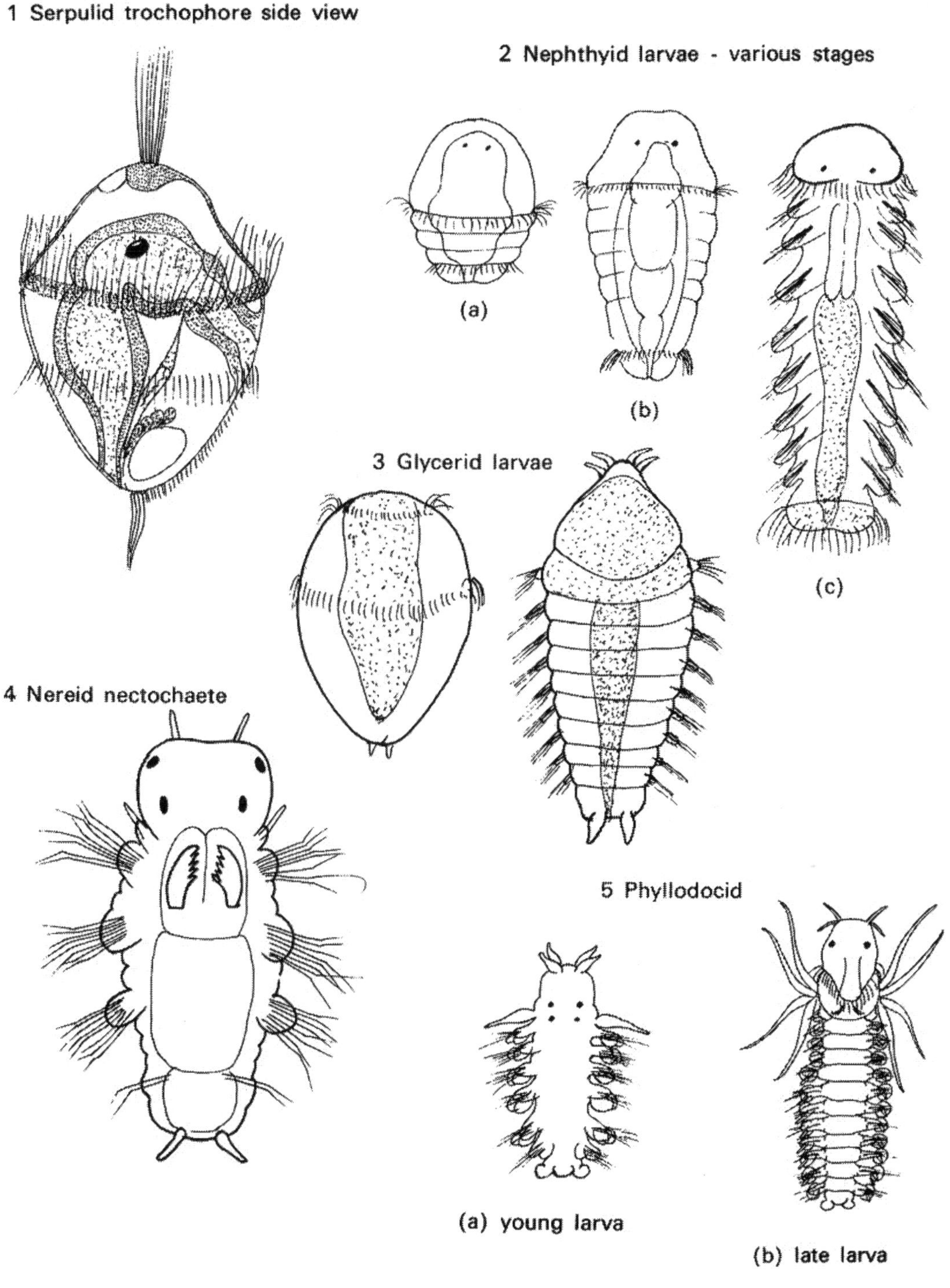

Plate 38. Polychaete larvae

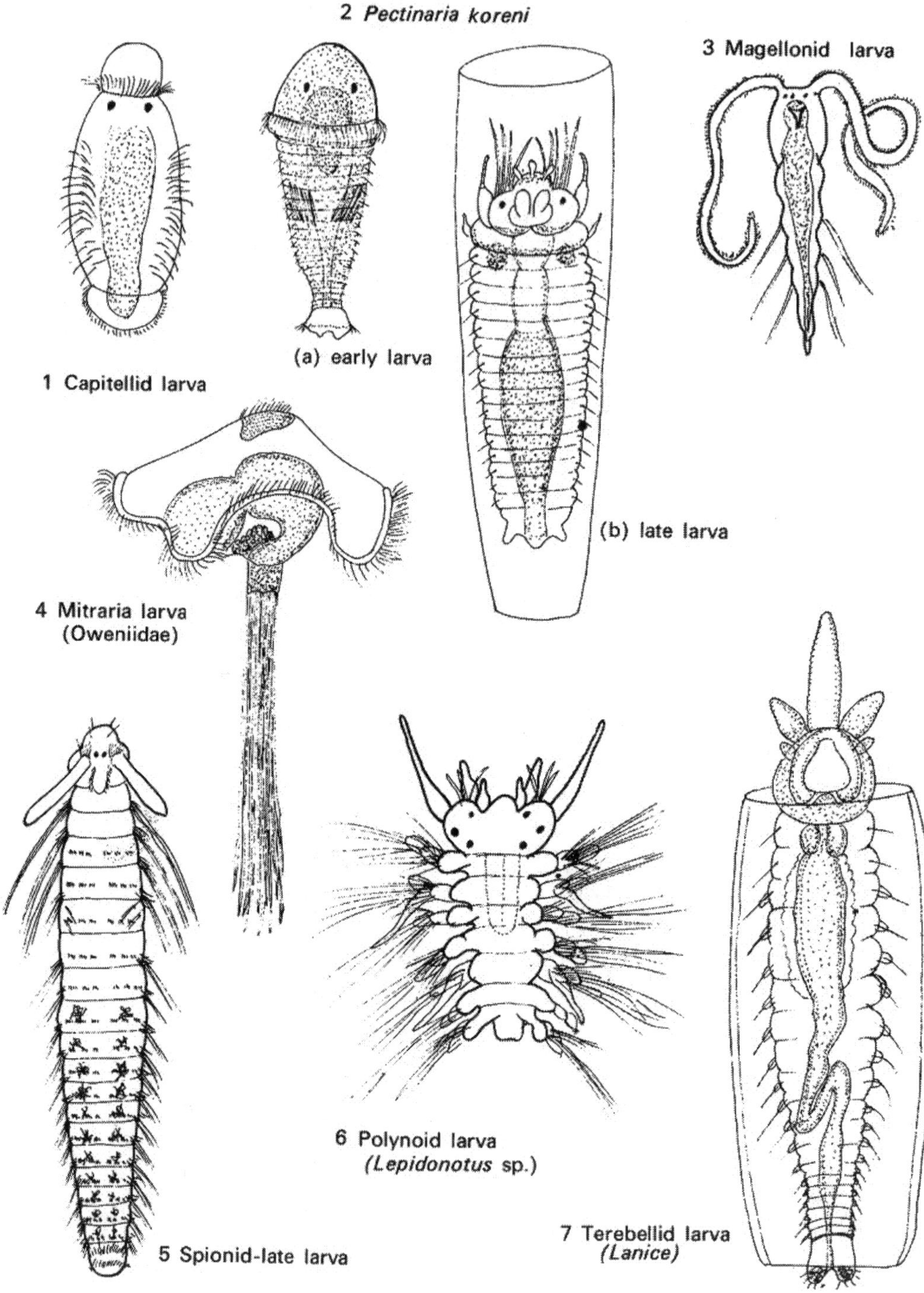

Plate 39. Polychaete larvae

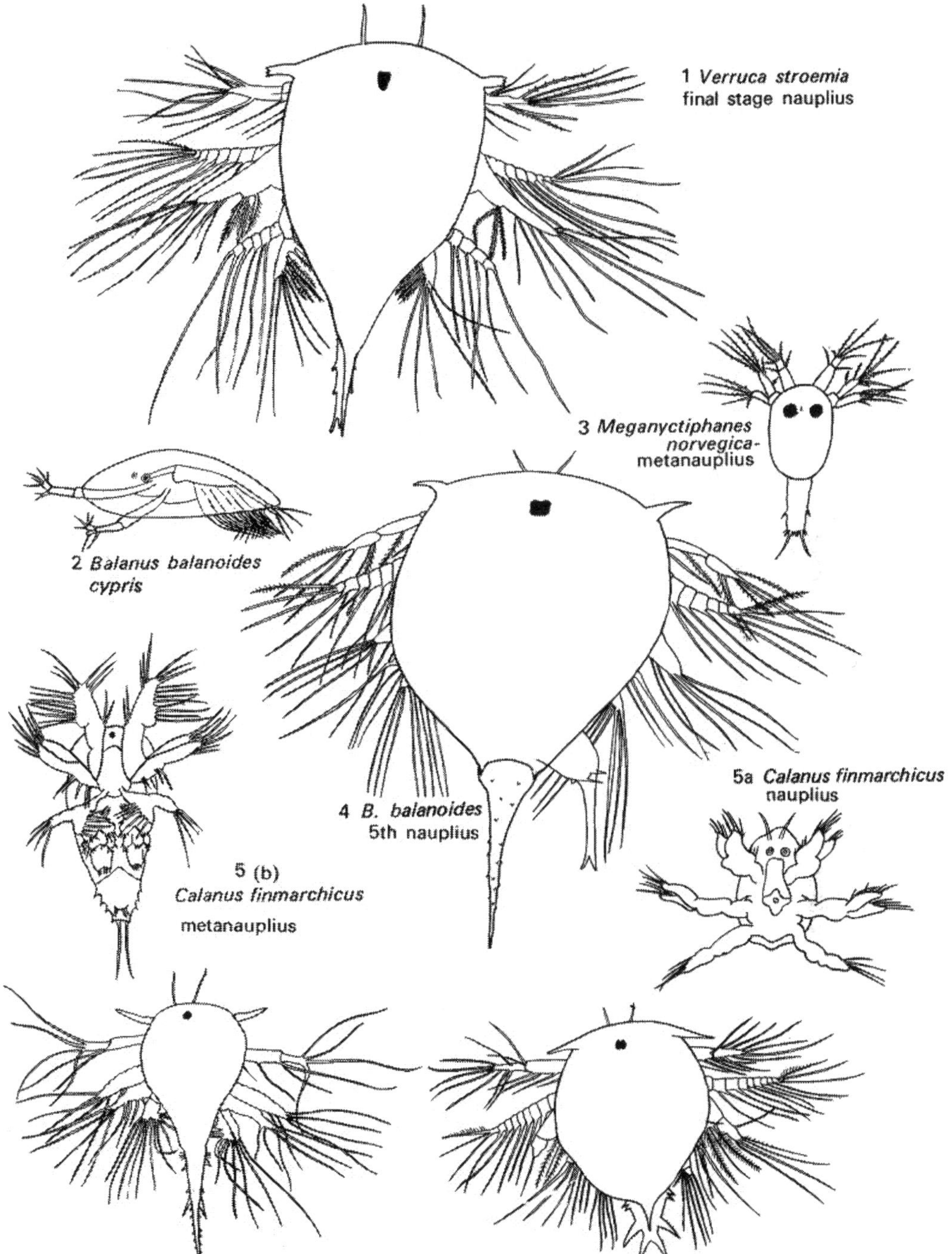

Plate 40. Nauplii

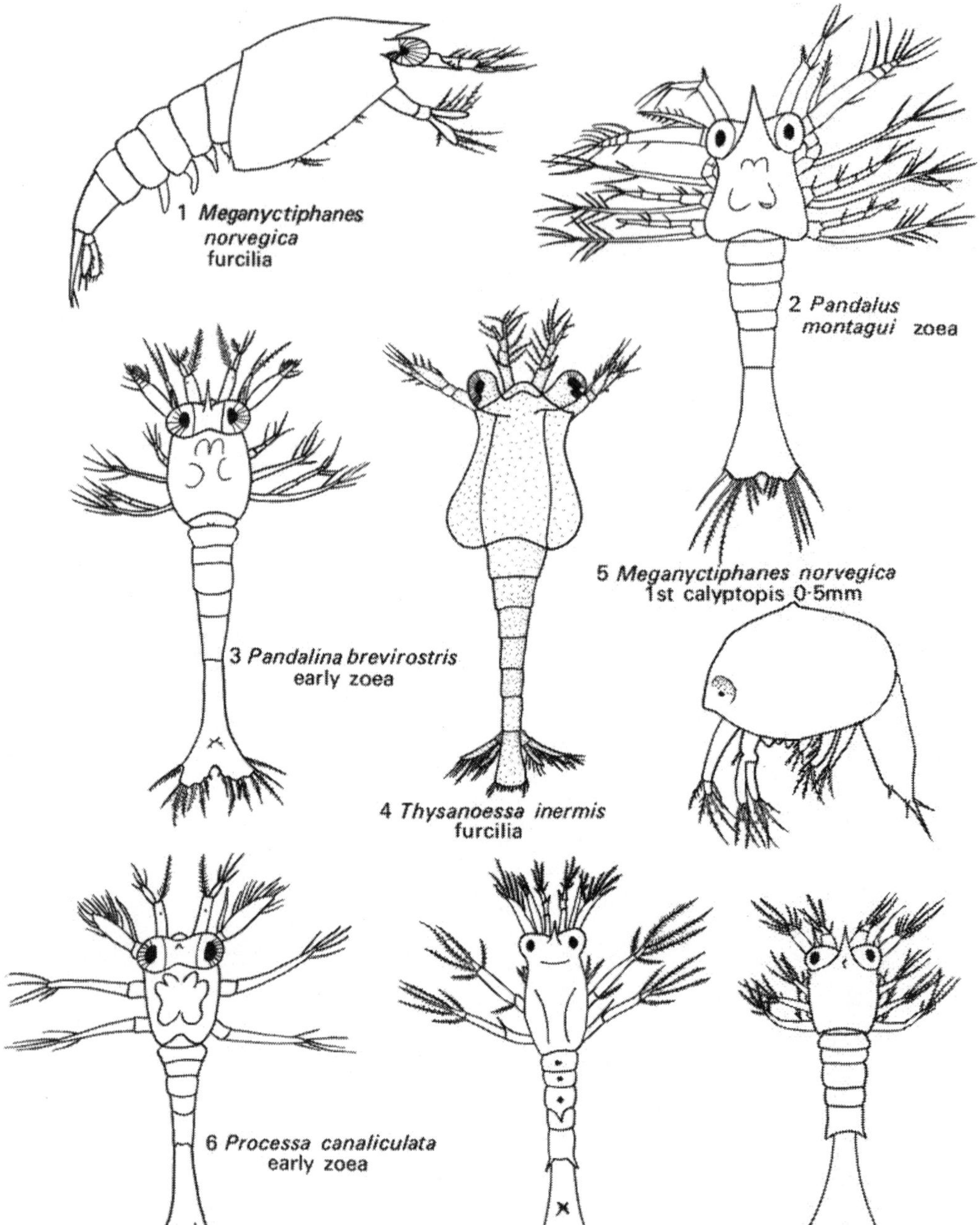

Plate 41. Euphausid and Caridean larvae

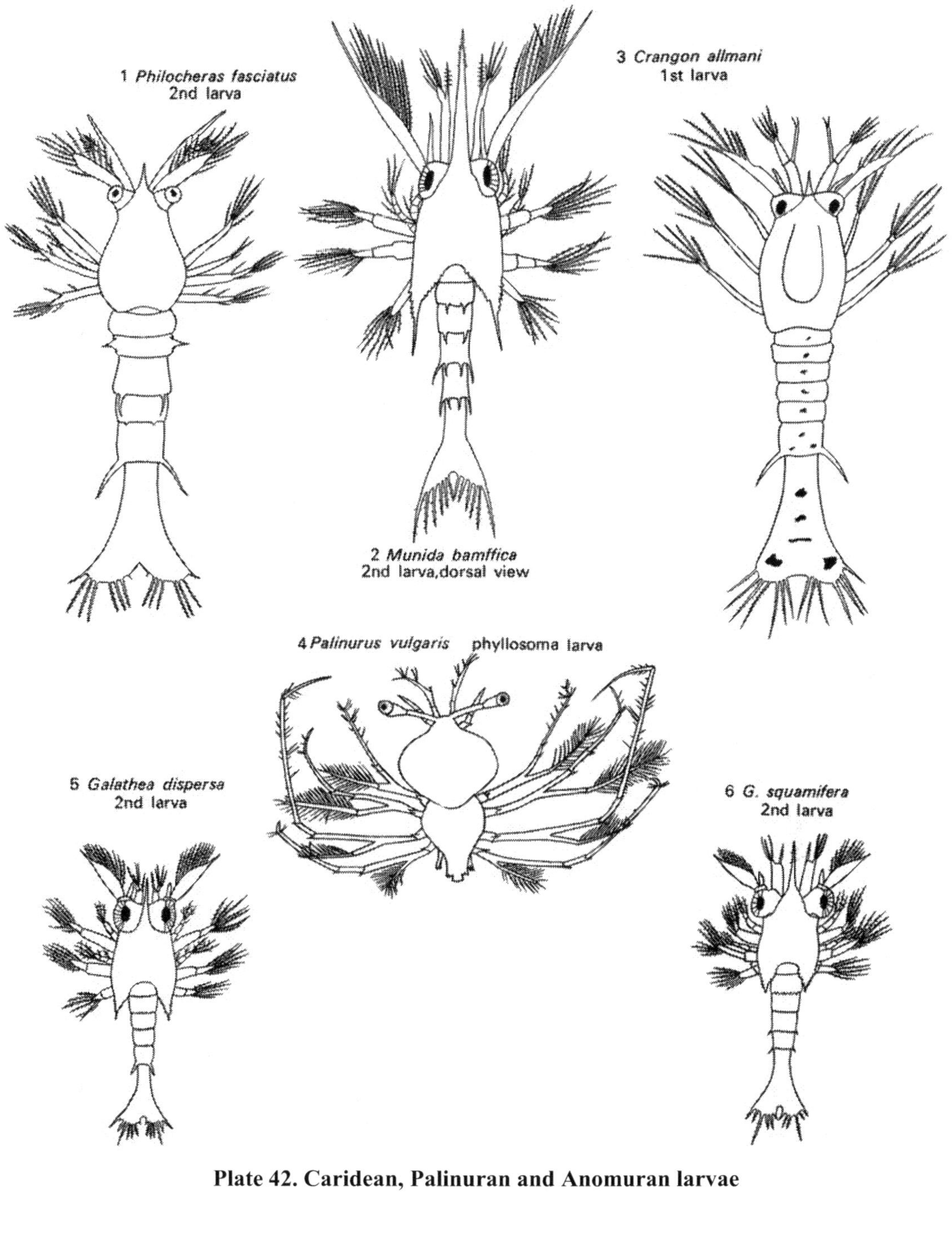

Plate 42. Caridean, Palinuran and Anomuran larvae

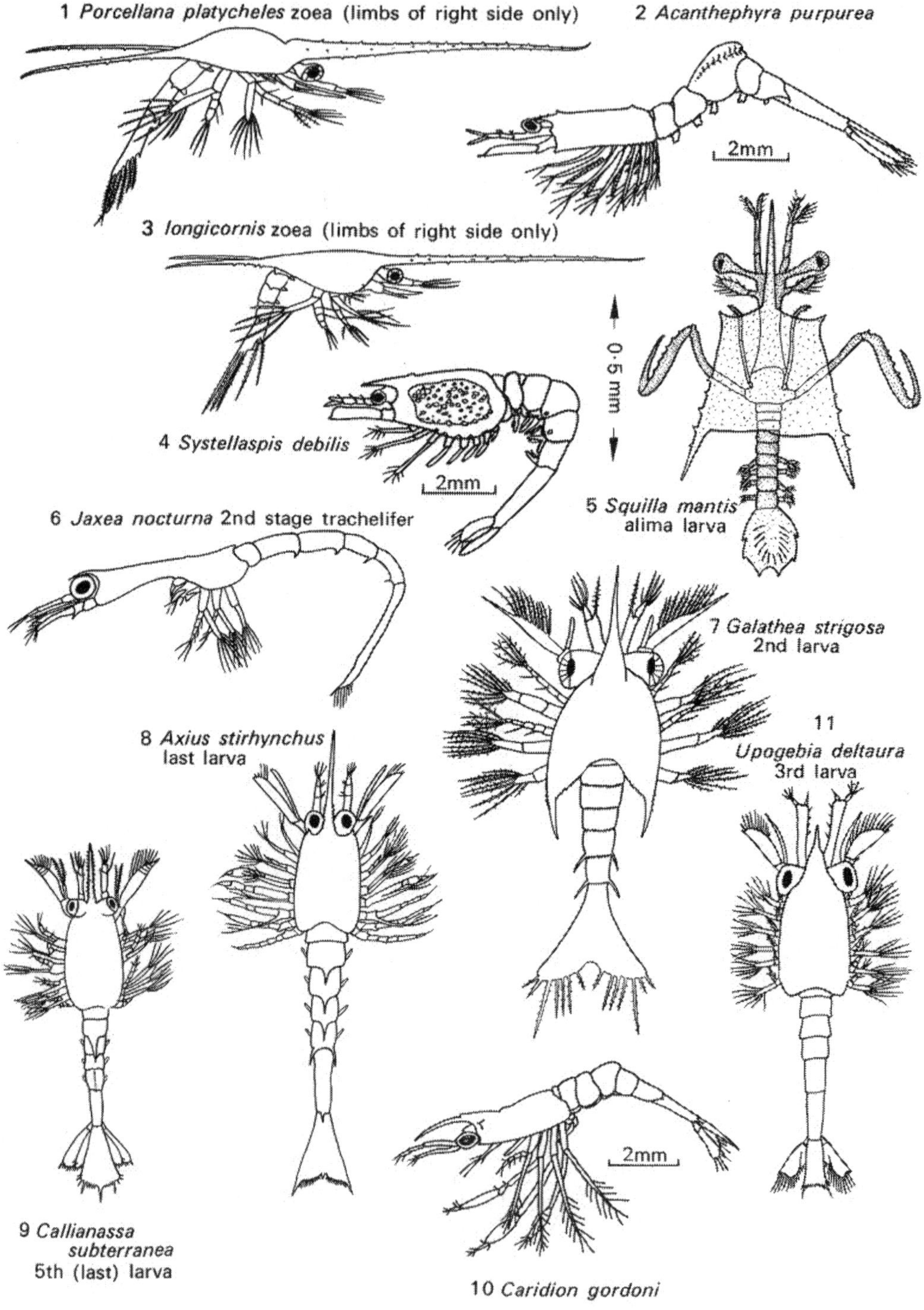

Plate 43. Anomuran Larvae, *Squilla* Larva

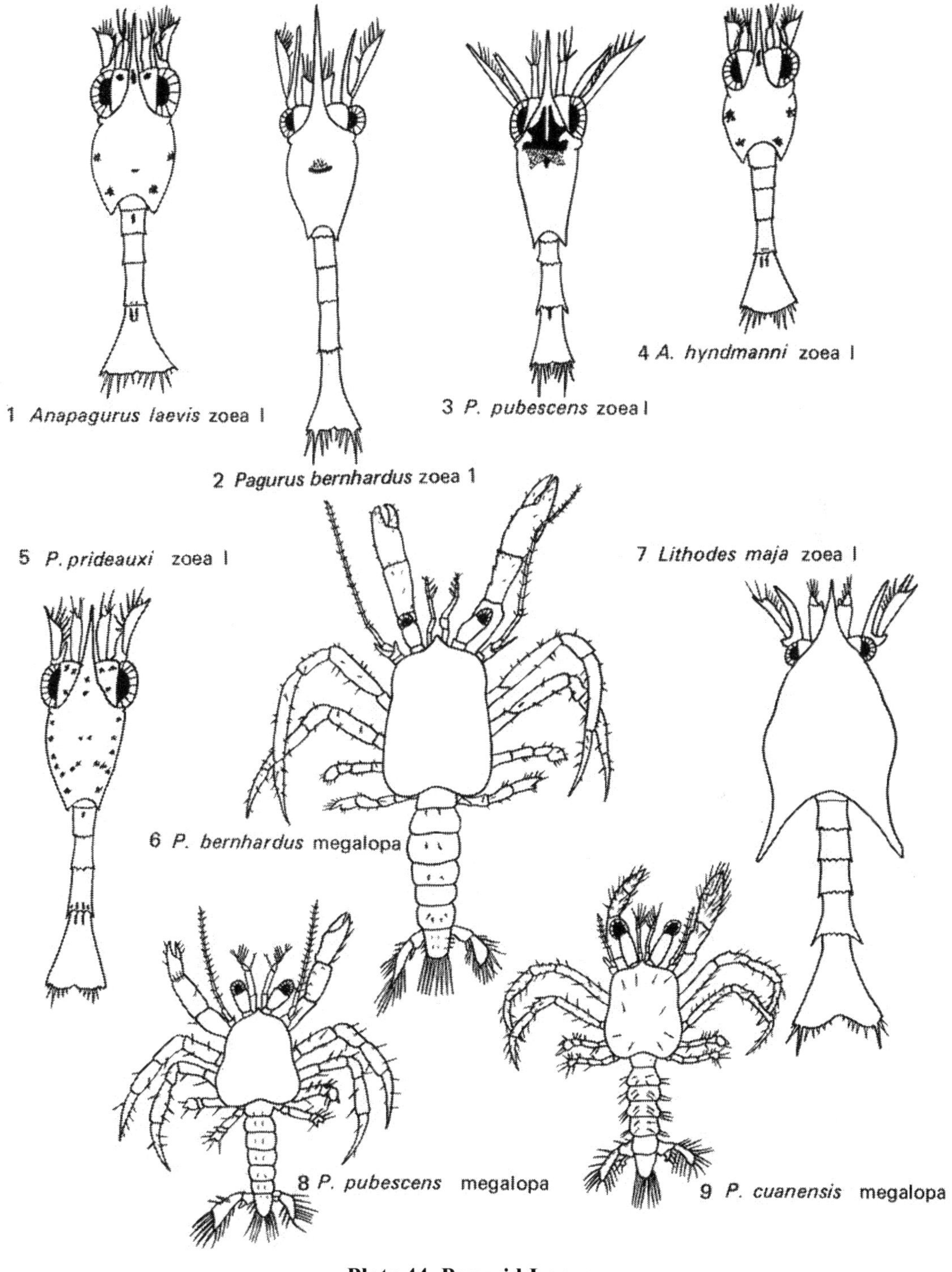

Plate 44. Pagurid Larvae

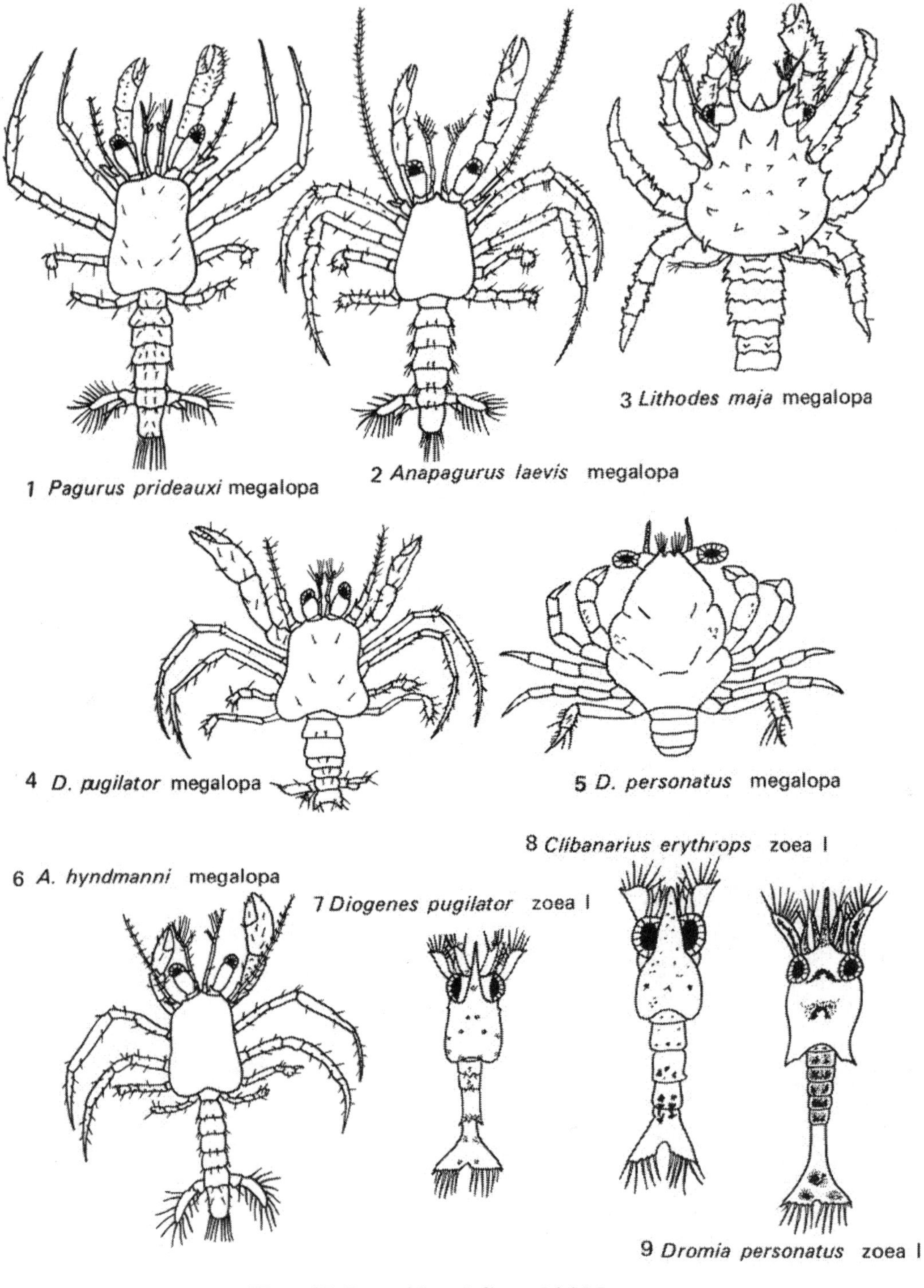

Plate 45. Pagurid and Coenobitid Larvae

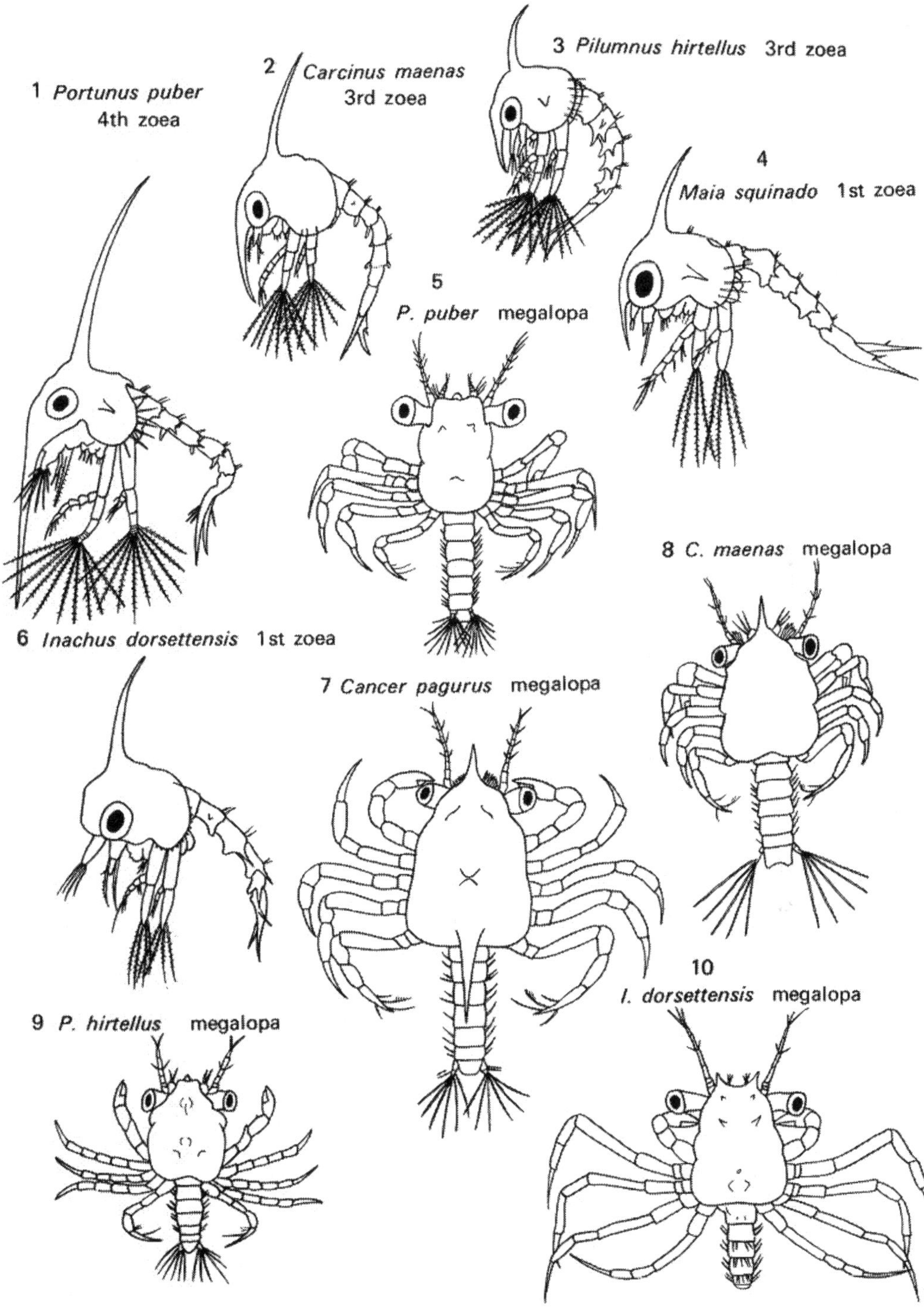

Plate 46. Brachyuran Larvae

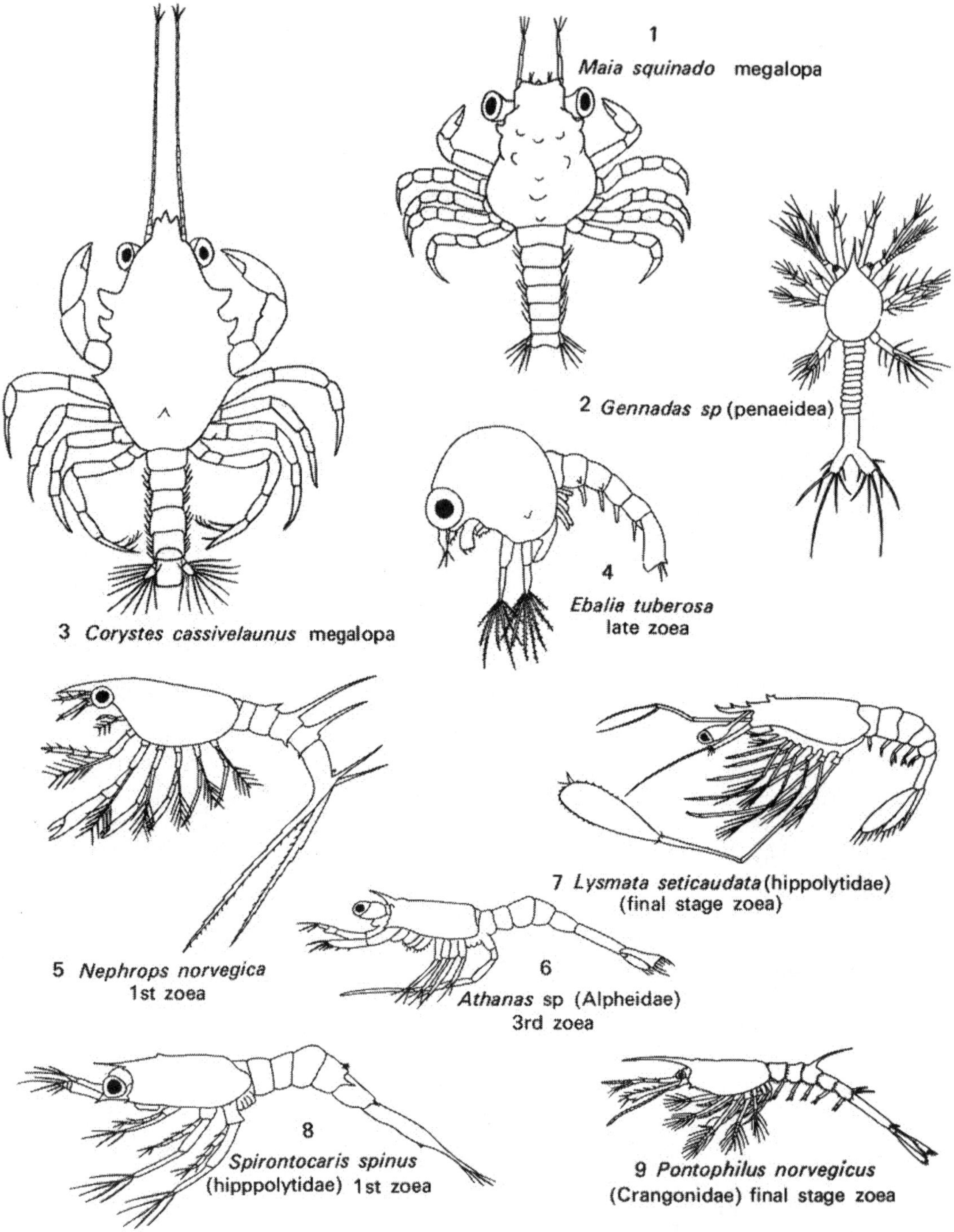

Plate 47. Megalopas and zoeas

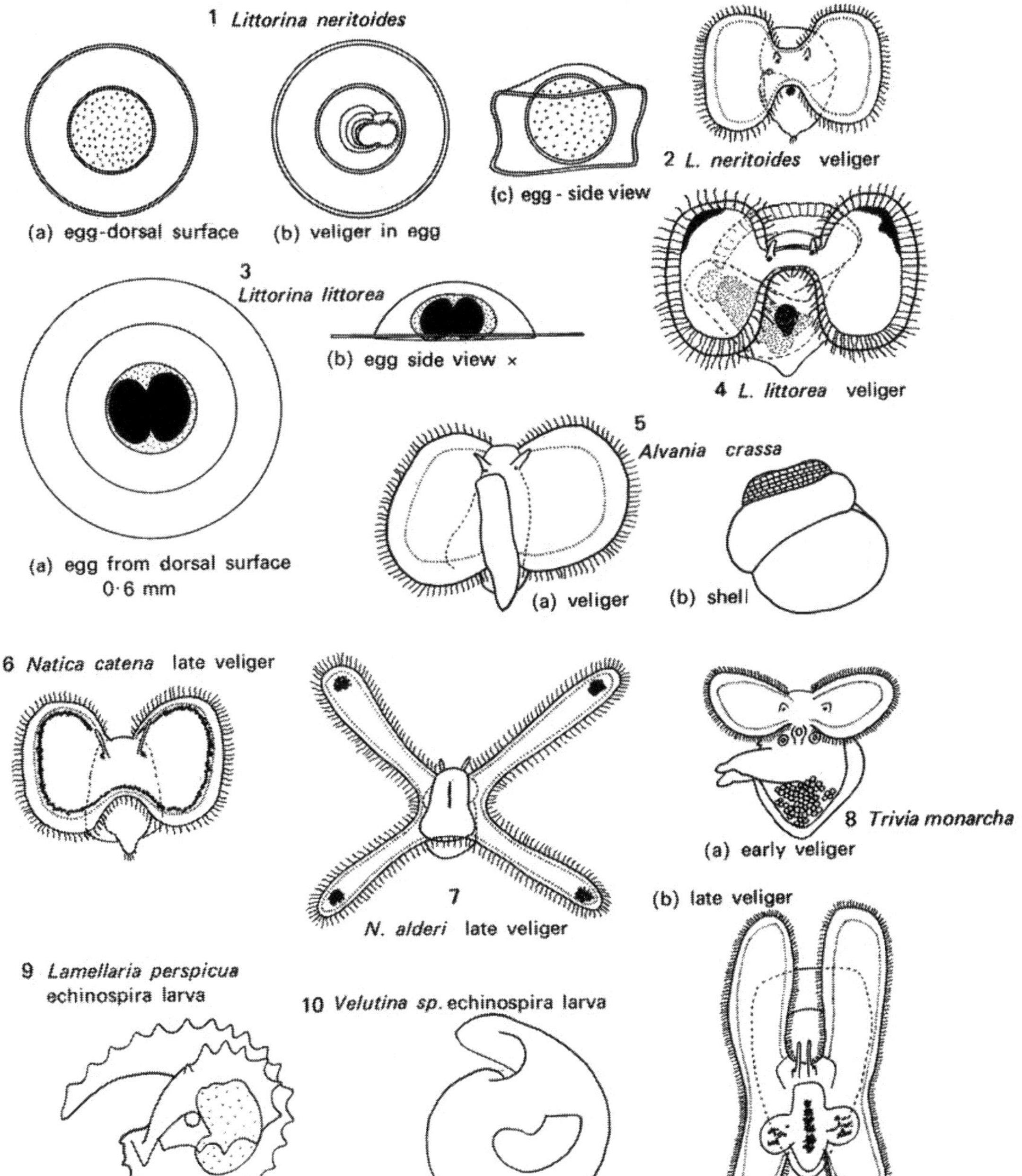

Plate 48. Gastropod Larvae

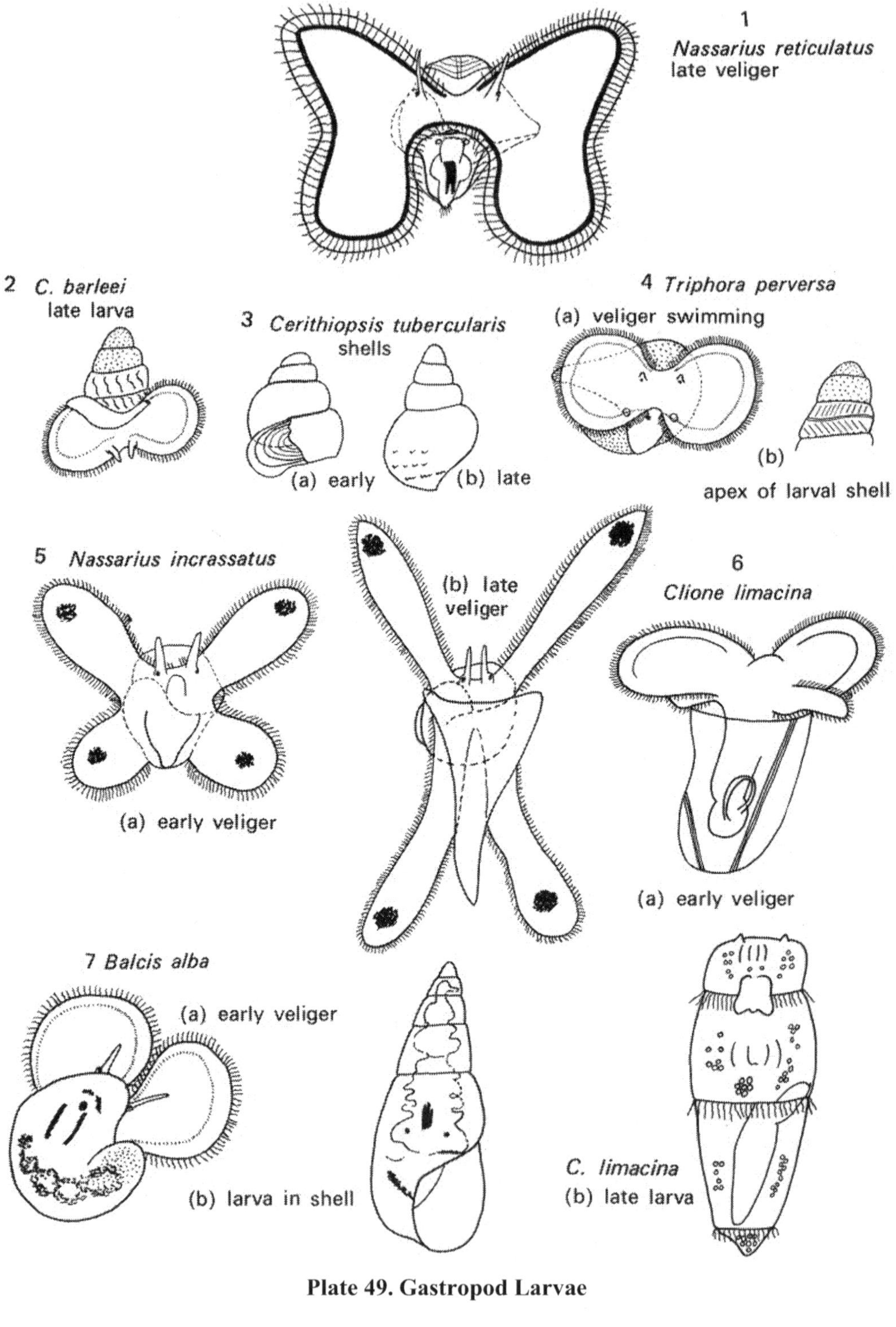

Plate 49. Gastropod Larvae

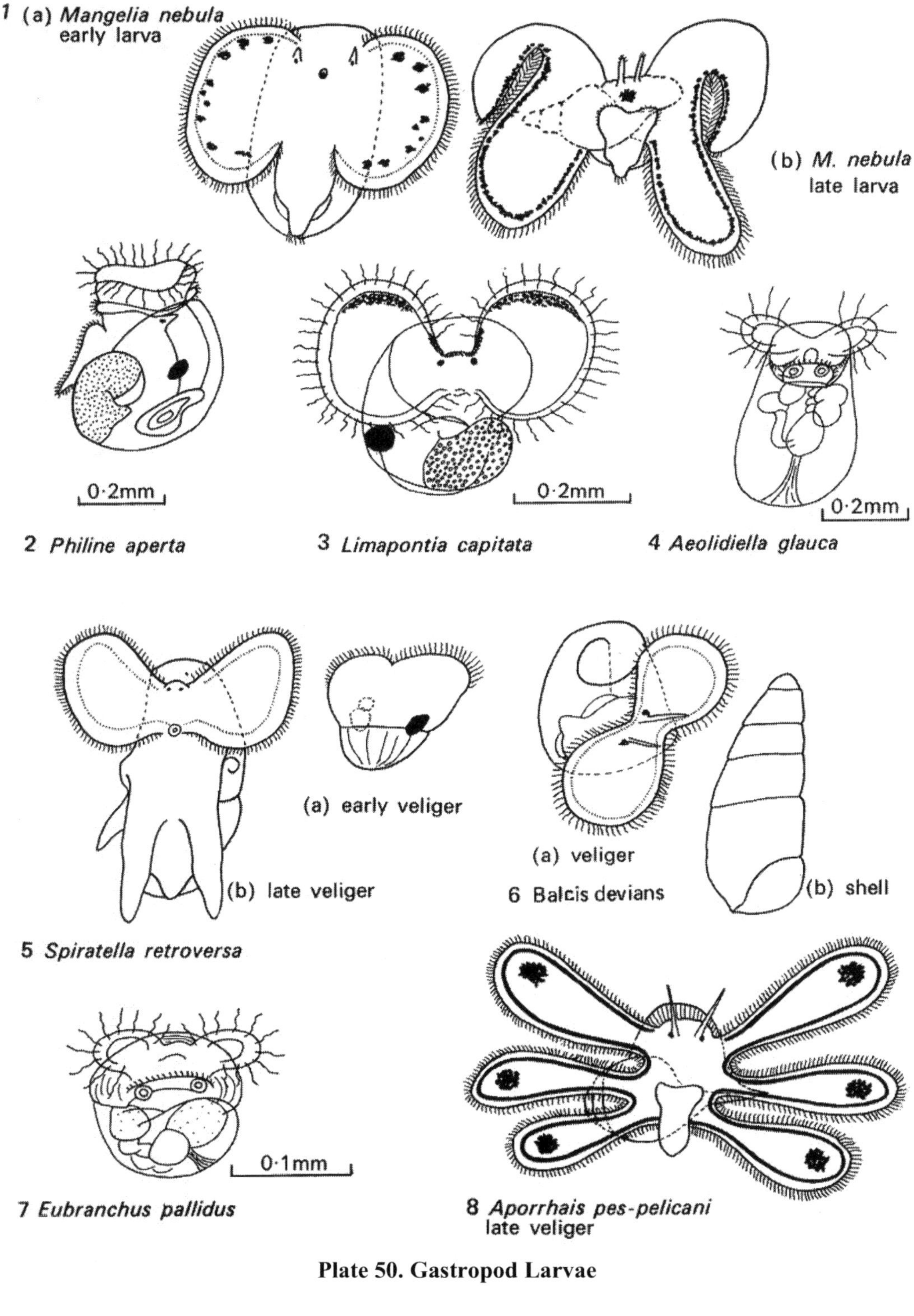

Plate 50. Gastropod Larvae

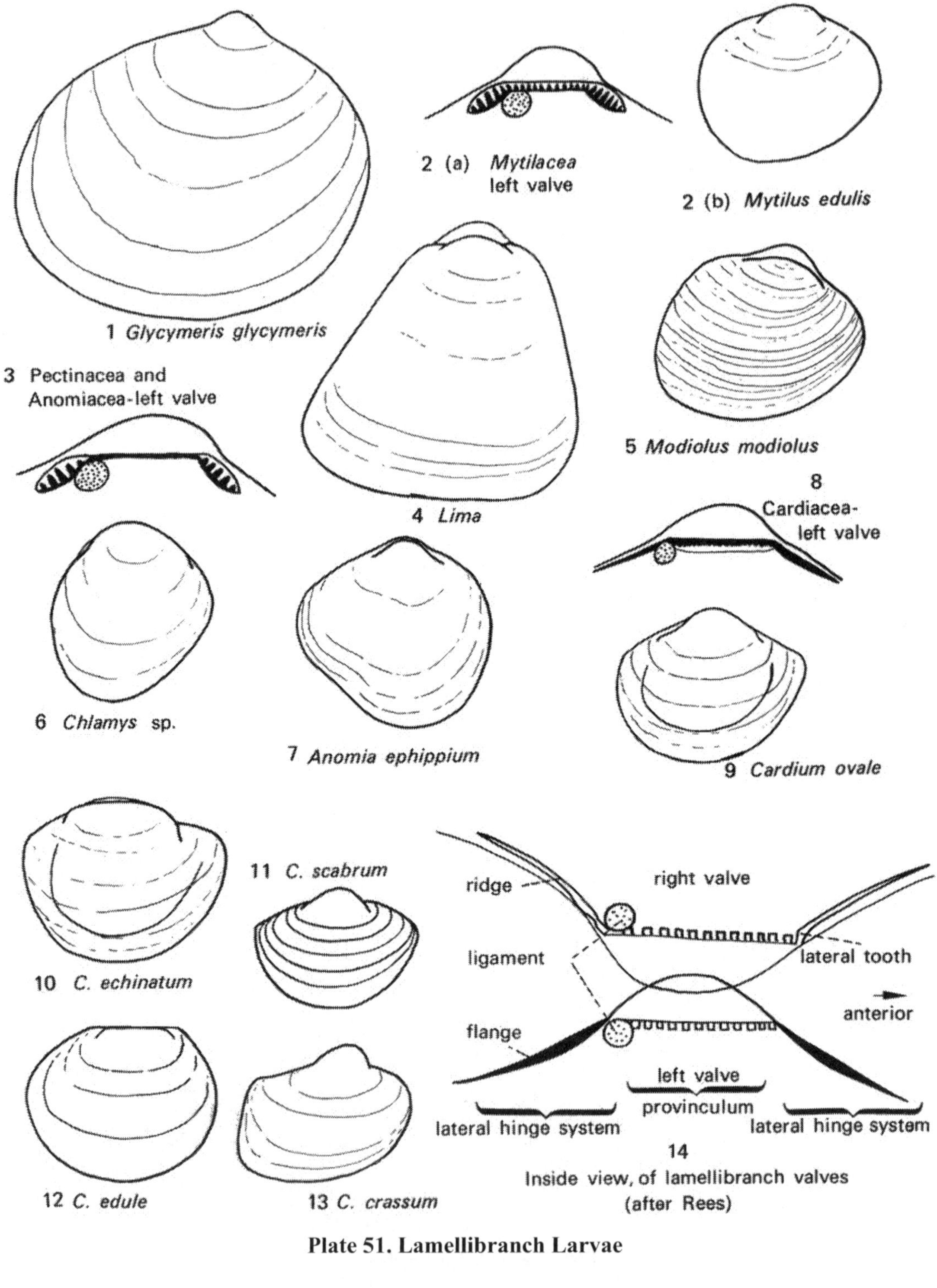

Plate 51. Lamellibranch Larvae

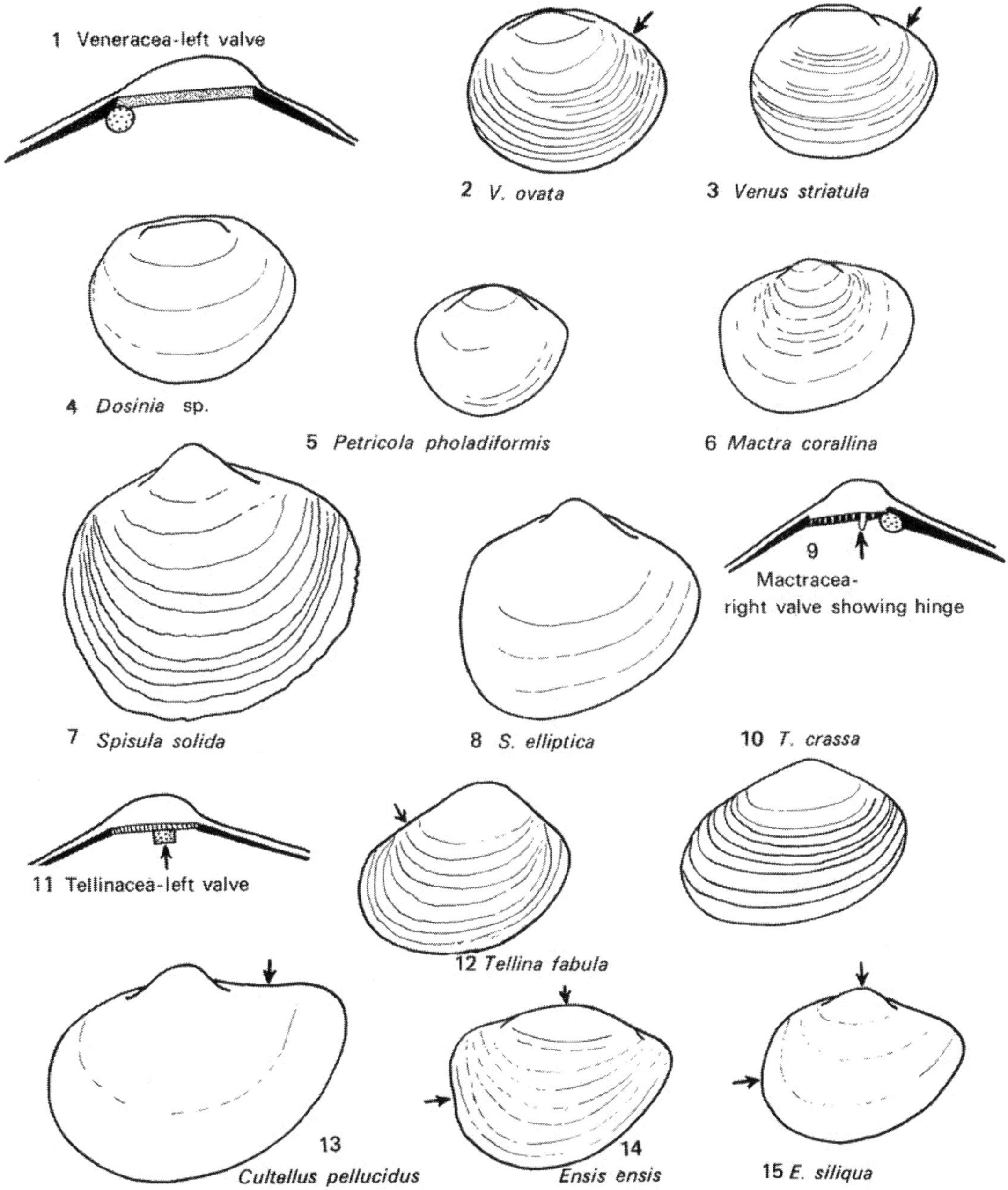

Plate 52. Lamellibranch Larvae

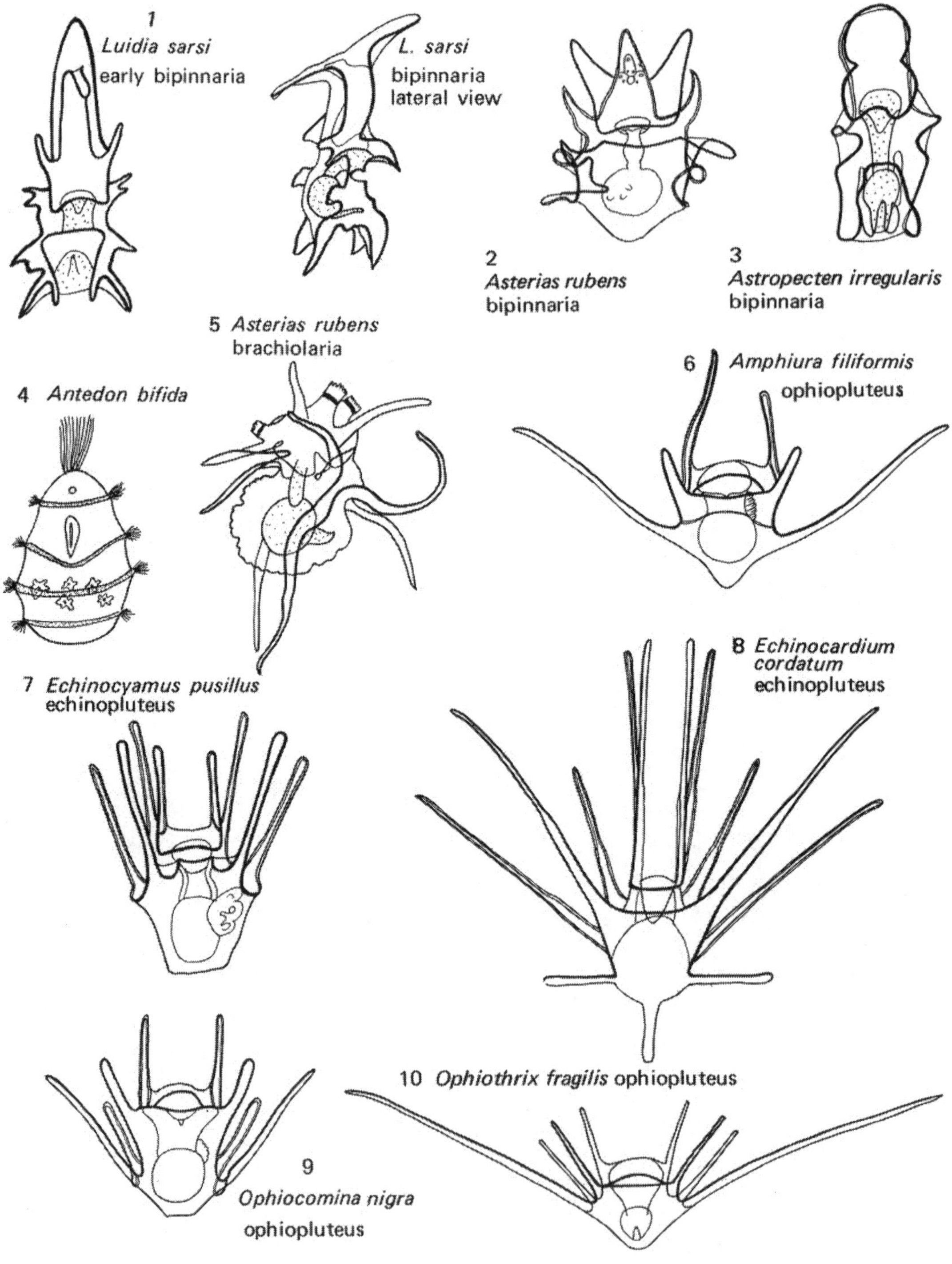

Plate 53. Echinoderm Larvae

1 *Psammechinus miliaris*
echinopluteus

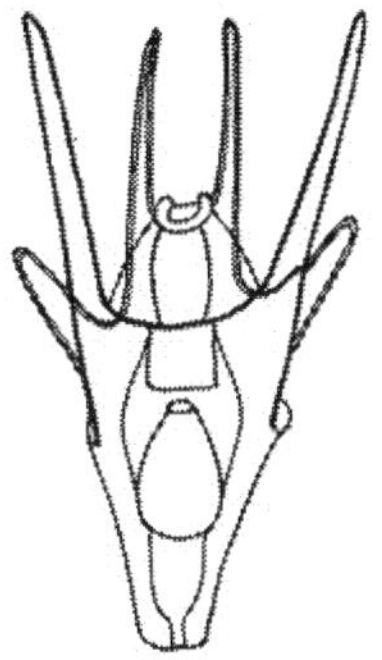

2 *Spatangus purpureus*
young echinopluteus

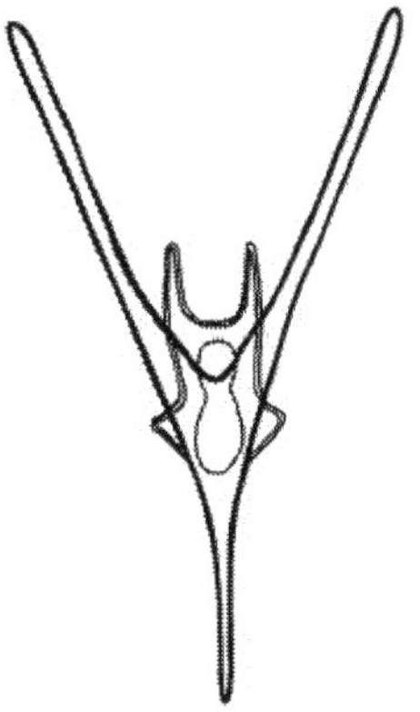

3 *Labidoplax digitata*
auricularia

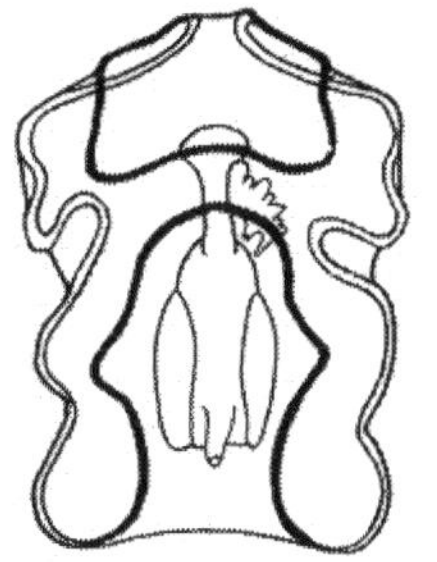

4 *Cucumaria sp*
barrel-larva

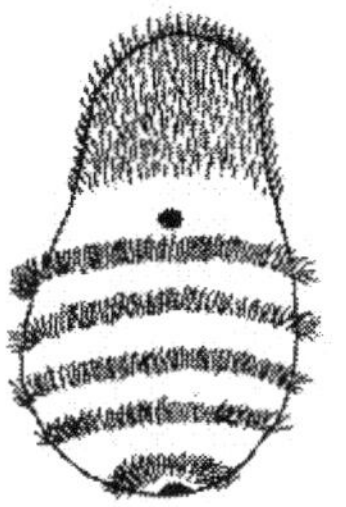

5 *Holothuria sp.*
pentacula larva

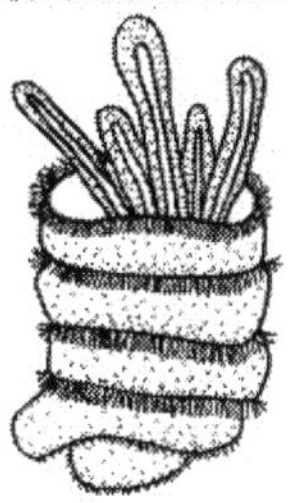

6 *Leptosynapta inhaerens*
barrel-larva

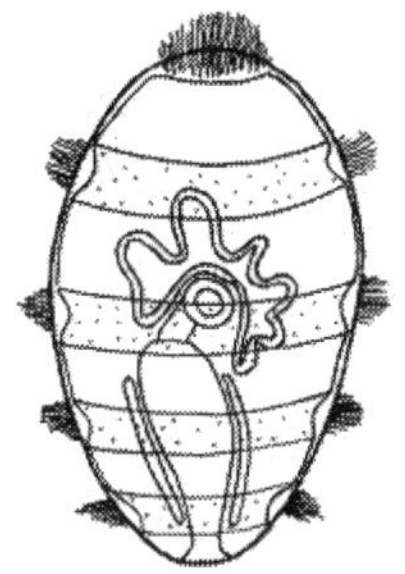

7 *Balanoglossus clavigerus*
tornaria larva

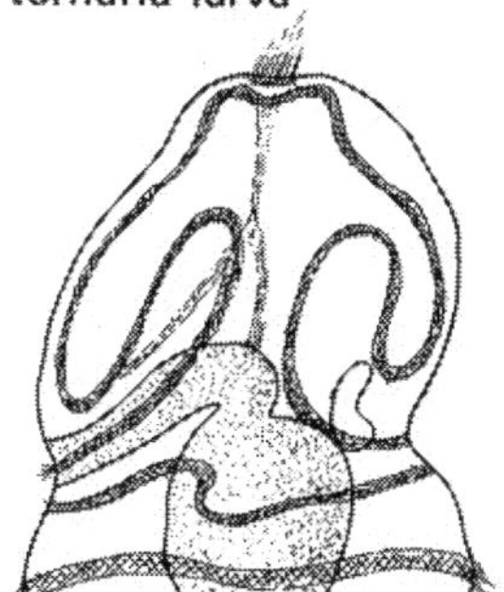

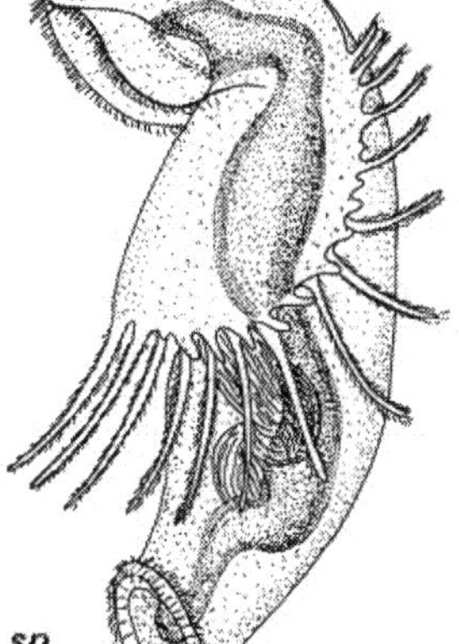

8 *Phoronis sp.*
actinotrocha

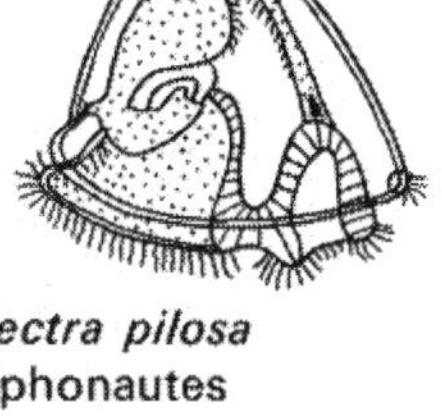

9 *Electra pilosa*
cyphonautes

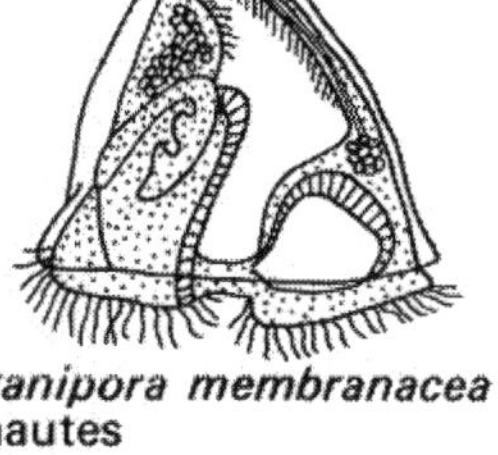

10 *Membranipora membranacea*
cyphonautes

Plate 54. Larvae of Echinoderms, Ectoprocts, Enteropneusts and *Phoronis*

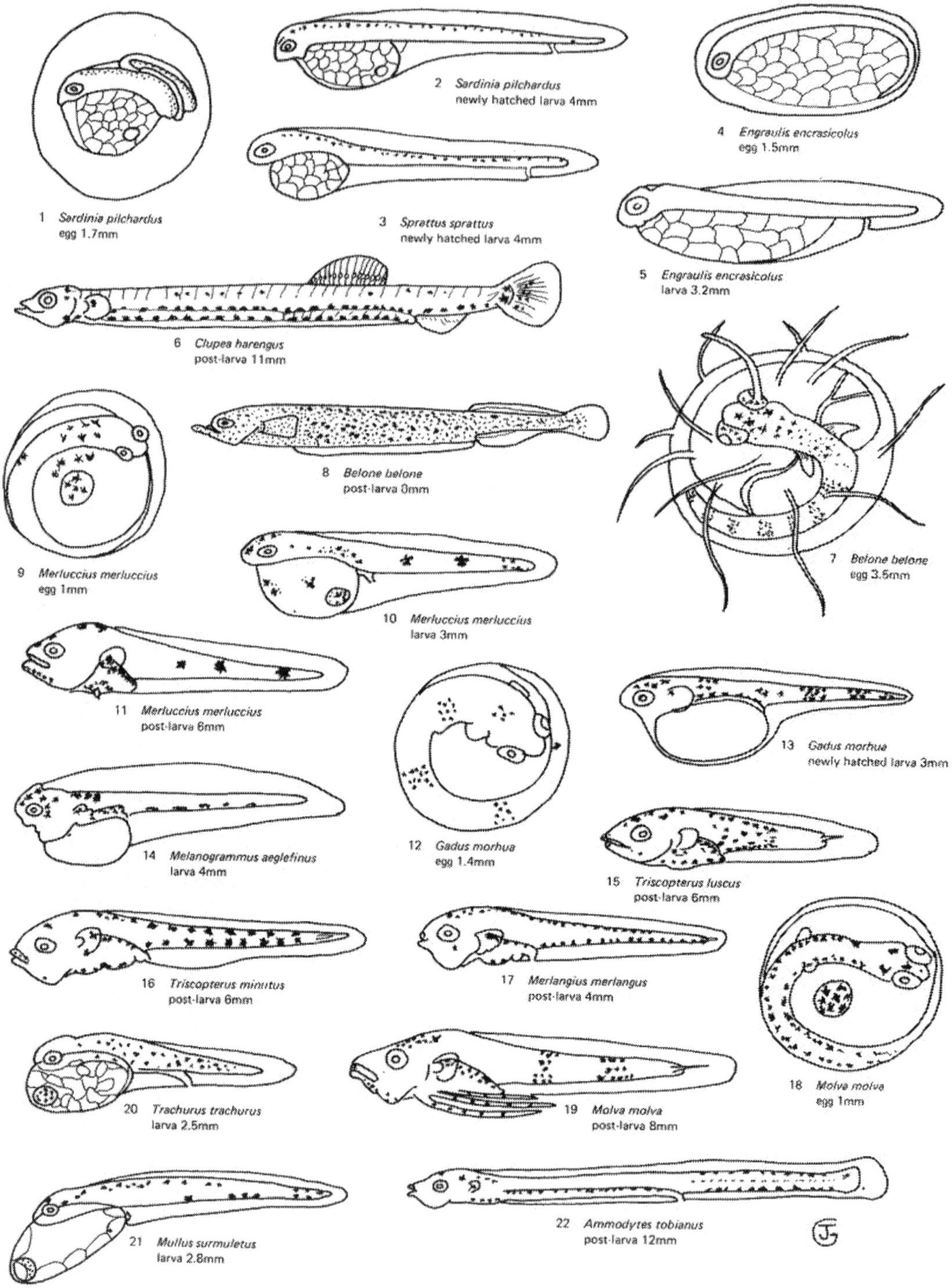

Plate 55. Fish Eggs and Larvae

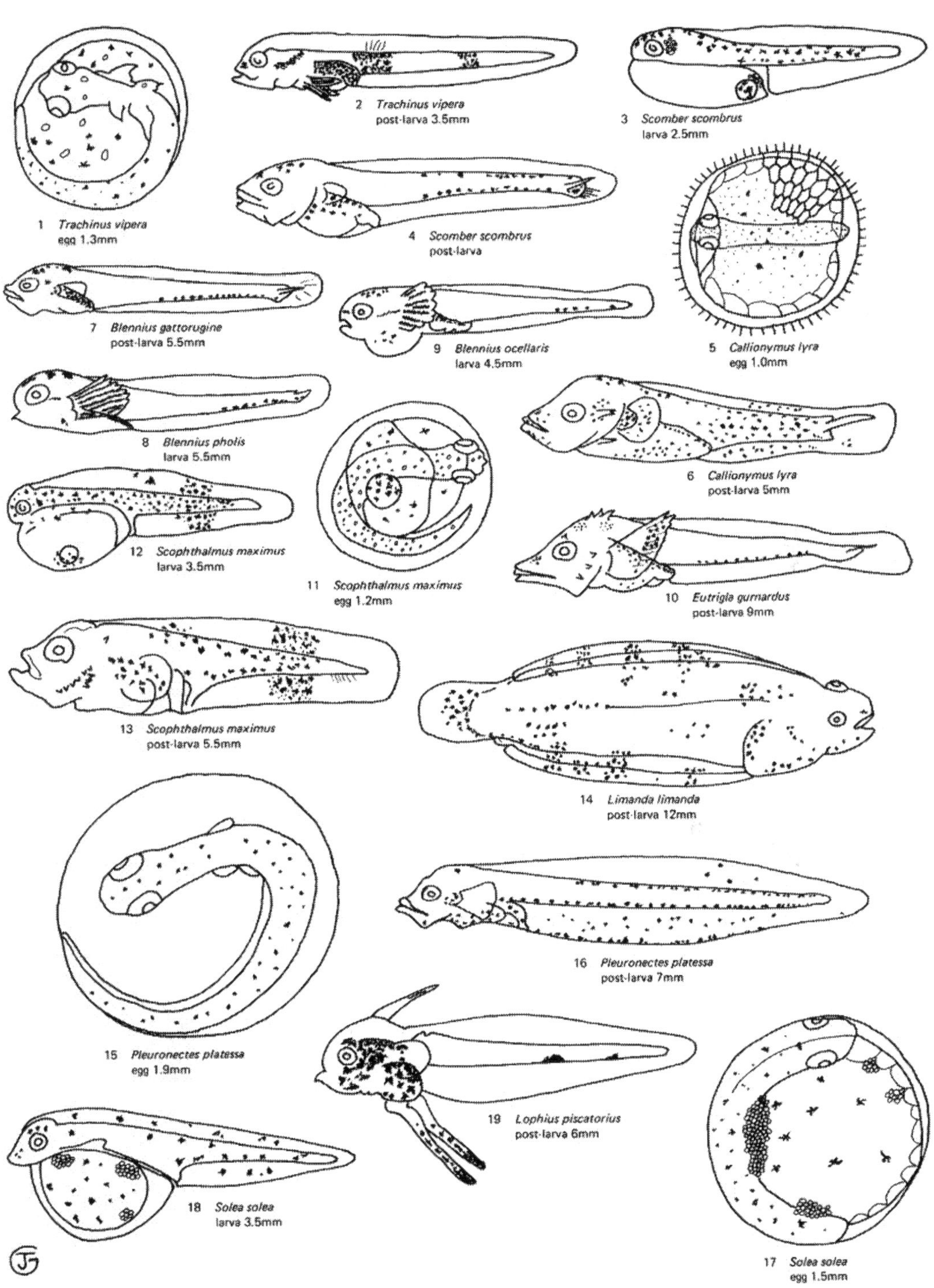

Plate 56. Fish Eggs and Larvae

References to General Topics

ARMSTRONG, F. A. J. and WICKSTEAD, J. H. 1962. A note on the preservation of plankton samples with formalin: J. Cons. int. Explor. Mer. **27**, 29-30

ARON, W. 1958. The use of a large capacity portable pump for plankton sampling, with notes on plankton patchiness: J. Mar. Res. **16**, 158-73

ARON, W. 1961. Some aspects of sampling the macroplankton. Symposium on zooplankton production: Copenhagen, 1961, ed. J. H. F. and J. Corlett, Rapp. Cons. Explor. Mer. **153**, 232 pp.

BANSE, K. 1955. Über das Verhalten von meroplanktischen Larven in geschichtetem Wasser: Kiele Meeresforsch. **11**, 188-200

BANSE, K. 1956. Über den Transport von meroplanktischen Larven aus dem Kattegat in die Kieler Bucht: Ber. Dtsch. Komm. Meeresforsch. **14,** 147-64

BANSE, K. 1959. Die Vertikalverteilung planktischer Copepoden in der Kieler Bucht: Ber. Dtsch. Komm. Meeresforsch. **15,** 357-88

BARNES, H. 1949*a*. On the volume measurement of water filtered by a plankton pump, with some observations on the distribution of planktonic animals: J. mar. biol. Ass. U.K. **28**, 651-62

BARNES, H. 1949*b*. A statistical study of the variation in vertical plankton hauls, with special reference to the loss of catch with divided hauls: J. mar. biol. Ass. U.K. **28**, 428-46

BARNES, H. 1958. Oceanography and marine biology: 218 pp. Allen & Unwin, London

CARRUTHERS, J. N. 1935. The flow of water through the Strait of Dover as gauged by continuous current-meter observations at the Varne Lightvessel (50° 56'N-1° 17'E.). Part II. Second report on results obtained: Fish. Invest. Ser. 2 **14,** No. 4, 1-67

CLARK, R. B. 1953. Pelagic swimming of Scalibregmidae (Polychaeta): Rep. Scot. mar. biol. Ass. 20-2

CLARK, R. S. 1940. The pelagic young and early bottom stages of teleosteans: J. mar. biol. Ass. U.K. **12,** 159-240

CLARKE, G. L. 1946. The dynamics of production in a marine area: Ecol. Monographs. **16,** 323-5

CLARKE, G. L. and BUMPUS, D. F. 1950. The plankton sampler - an instrument for quantitative plankton investigations: Amer. Soc. Limnol. Oceanogr. Special Publication No. 5, 8 pp.

COLE, H. A. and KNIGHT-JONES, E. W. 1939. Some observations on the setting behaviour of larvae of *Ostrea edulis*: J. Cons. int. Explor. Mer. **14,** 86-105

COLMAN, J. S. and SEGROVE, S. 1955. The tidal plankton over Stoupe Beck sand, Robin Hood's Bay (Yorkshire, North Riding): J. Anim. Ecol. **24,** 445-62

COOPER, L. H. N. 1952. The physical and chemical oceanography of the waters bathing the continental slope of the Celtic Sea: J. mar. biol. Ass. U.K. **30**, 465-510

CURRIE, R. I. 1961. Net closing gear. Symposium on zooplankton production: Copenhagen 1961, ed. J. H. F. and J. Corlett: Rapp. Cons. Explor. Mer. **153**, 232 pp.

CURRIE, R. I. and FOXTON, P. 1957. A new quantitative plankton net: J. mar. biol. Ass. U.K. **36**, 17-32

DE BLOK, J. W. and GEELEN, H. J. F. M. 1958. The substratum required for the settling of mussels (*Mytilus edulis L.*): Arch. Neerl. Zool. **13,** 446-60

EVANS, F. and NEWELL, G. E. 1957. Tidal streams and larval dispersal at Whitstable: Ann. Mag. nat. Hist. Ser. 12. **10,** 161-73

FAGE, L. and LEGENDRE, R. 1927. Pèches planktoniques à la lumière effectuées a Banyuls-sur-Mer et a Concarneau: Arch. Zool. exp. gén. **67**, 23-222

FISCHER PIETTE, E. 1936. Etudes sur la biogéographie intercòtidale des deux rives de la Manche. J. Linn. Soc. **40**, 181-272

FRASER, J. H. 1952. The Chaetognatha and other zooplankton of the Scottish area and their value as biological indicators of hydrological conditions: Mar. Res. Scot. 1952

FRASER, J. H. 1955. The plankton of the waters approaching the British Isles in 1953: Mar. Res. Scotland for 1955

FRASER. J. H. 1962. Nature adrift, the story of marine plankton: 178 pp., Foulis, London

FRASER, J. H. 1965, Zooplankton indicator species in the North Sea. In: Serial Atlas of the Marine Environment, Folio 8. Amer. Geograph. Soc.

FRASER, J. H. 1966. Zooplankton Sampling. Nature Lond. **211,** No. 5052, 915-16

GEHRINGER, J. W. 1961. The Gulf III and other modern high-speed plankton samplers. Symposium on zooplankton production: Copenhagen, 1961, ed. J. H. F. and J. Corlett, Rapp. Cons. Explor. Mer. **153**, 232 pp.

GLOVER, R. S. 1953. The Hardy plankton indicator and sampler: a description of the various models in use: Bull. Mar. Ecol. **4**, 7-20

GLOVER, R. S. 1961. The multi-depth plankton indicator: Bull. Mar. Ecol. **V**, 151-64

HANSEN, V. G. and ANDERSON, K. D. 1961. Sampling the smaller zooplankton. Symposium on zooplankton production: Copenhagen, 1961, ed. J. H. F. and J. Corlett, Rapp. Cons. Explor. Mer. **153**, 232 pp.

HARDY, A. C. 1926. The herring in relation to its animate environment. Part II. Report on trials with the plankton indicators: Fish. Invest. Ser. II, **7**, No. 7, 1-13

HARDY, A. C. 1936. The continuous plankton recorder, with appendix: test of validity of the continuous plankton recorder method with Ennis, N.: Discovery Rept. **11,** 457-510

HARDY, A. C. 1939. Ecological investigations with the continuous plankton recorder: Bull. mar. Ecol. **1**, No. 1, 1-57

HARDY, A. C. 1956. The open sea. Its natural history; the world of plankton: 336 pp. New Naturalist series. Collins, London

HARDY, A. C. and BAINBRIDGE, R. 1954. Experimental observations on the vertical migrations of plankton animals: J. mar. biol. Ass. U.K. **33**, 409-48

HARVEY, H. W. 1934. Measurement of phytoplankton population: J. mar. biol. Ass. U.K. **19**, 761-73

HARVEY, H. W. 1942. Production of life in the sea: Biol. Rev. 17, 221-46

ISAACS, J. D. and KIDD, L. W. 1953. The Isaacs-Kidd midwater trawl: Scripps. Inst. Ocean, **53-3**, 18 pp.

KNIGHT-JONES, E. W. 1951. Gregariousness and some other aspects of the setting behaviour of *Spirorbis borealis* (Serpulidae): J. mar. biol. Ass. U.K. **30**, 201-22

KNIGHT-JONES, E. W. 1953a. Laboratory experiments on gregariousness during setting in *Balanus balanoides* and other barnacles: J. expt. Biol. **30**, 584-598

KNIGHT-JONES, E. W. 1953b. Some further observations on gregariousness in marine larvae:

Brit. J. anim. Behav. **1**, 81-2
KNIGHT-JONES, E. W. and CRISP, D. J. 1953. Gregariousness in barnacles in relation to the fouling of ships and to anti-fouling research: Nature, Lond. **171**, 1109-10
KORRINGA, P. 1957. Lunar periodicity *in* Treatise on marine ecology and paleoecology. **I**, ed. Hedgpeth, J. W. Mem. 67. Geol. Soc. Amer. 917-34
MAGHRABY, A. M. el. 1955. A qualitative and quantitative survey of the plankton of the Thames estuary, Whitstable: Ph.D. thesis, University of London
MILLER, D. 1961. A modification of the small Hardy plankton sampler for simultaneous high-speed plankton hauls: Bull. Mar. Ecol. **V**, 165-72
NEWELL, G. E. 1948. A contribution to our knowledge of the life-history of *Arenicola marina* L.: J. mar. biol. Ass. U.K. **27**, 554-80
NEWELL, G. E. 1949a. The later larval life of *Arenicola marina.* (*L*).: J. mar. biol. Ass. U.K. **28**, 635-39
NEWELL, G. E. 1949b. *Clymenella torquata* (Leidy), a polychaete new to Britain: Ann. Mag. nat. Hist. ser. 12, **2**, 147-56
NEWELL, G. E. 1951. The life-history of *Clymenella torquata* (Leidy): Proc. zool. Soc. Lond, **121**, 561-86
NEWELL, G. E. 1959. Pollution and the abundance of animals in estuaries. 61-69 *in* The effects of pollution on living material: Symposia of the Institute of Biology, No. 8, ed. Yapp, W. B.
NEWELL, R. 1962. Behavioural aspects of the ecology of *Peringia* (=*Hydrobia*) *ulvae* (Pennant), Prosobranchia, Mesogastropoda: Proc. zool. Soc. London, **138**, 49-75
NYHOLM, K. G. 1950. Contributions to the life-history of the ampharetid, *Melinna cristata*: Zool. Bidrag. Uppsala, **29**, 79-91
PAQUETTE, R. G., SCOTT, E. L. and SUND, P. N. 1961. An enlarged Clarke-Bumpus plankton sampler: Limnol. Oceanogr. **6**, 230-3
PAQUETTE, R. G. and FROLANDER, H. F. 1957. Improvements in the Clarke-Bumpus plankton sampler: J. Cons. int. Explor. Mer. **22**, 284-8
RAE, K. M. 1956. Parameters of the marine environment: pp. 3-16 *in* Perspectives in marine biology, ed. Buzzati-Traverso. Univ. California Press
REES, C. B. 1939. The distribution of *Calanus finmarchicus* (Gunn) and its two forms in the North Sea, 1938-39: Bull. Mar. Ecol. **2**, No. 14, 215-75
RUSSELL, F. S. 1935. On the value of certain plankton animals as indicators of water movements in the English Channel and North Sea: J. mar. biol. Ass. U.K. **20**, 309-22
RUSSELL, F. S. 1936. The importance of certain plankton animals as indicators of water movements in the western end of the English Channel : Rapp. Cons. Explor. Mer. **100,** 7-10
RUSSELL, F. S. 1938. The Plymouth offshore medusa fauna.: J. mar. biol. Ass. U.K. **22**, 411-40
RUSSELL, F. S. 1939. Hydrographical and biological conditions in the North Sea as indicated by plankton organisms: J. Cons. int. Explor. Mer. **14**, 171-92
RUSSELL, F. S. 1952. The relation of plankton research to fisheries hydrography: Rapp. Cons. Explor. Mer. **131**, 28-34
RUSSELL, F. S. and COLMAN, J. E. 1931. Great Barrier Reef Expedition, 1928-9: Scientific Reports, 2, No. 2. The Zooplankton. I Gear, methods and station lists: 35 pp. British Mus. (Nat. Hist.)
RUSSELL, F. S. and Yonge, C. M. 1928 (3rd ed. due) The Seas: 392 pp. Warne, London
SEGROVE, F. 1941. The development of the serpulid. *Pomatoceros triqueter* L.: Quart. J. micr. Sci. **82**, 467-540
SHEARER, C. 1911. On the development and structure of the trochophore of *Hydroides*

uncinatus (*Eupomatus*): Quart. J. micr. Sci. **56,** 543-9

SMITH, J. E. 1953. Maintenance and spread of seashore faunas: Adv. Sci. Rep. Brit. Ass. **38**

SOUTHWARD, A. J. 1961. The distribution of some plankton animals in the English Channel and western approaches. I. Samples taken with stramin nets in 1955 and 1957: J. mar. biol. Ass. U.K. **41,** 17-35

SOUTHWARD, A. J. 1962. The distribution of some plankton animals in the English Channel and approaches. II. Surveys with the Gulf III high-speed sampler 1958-60: J. mar. biol. Ass. U.K. **42**, 275-375

SOUTHWARD, A. J. 1963. The distribution of some plankton animals in the English Channel and approaches. III Theories about long-term biological changes, including fish. J. mar. biol. Ass. U.K. **43,** 1-29.

STEELE, J. H. 1959. The quantitative ecology of marine phytoplankton: Biol. Rev. **34**, 129-58

STEEMAN NIELSEN, E. and JENSEN, E. A. 1957. Primary oceanic production: Galathea Rep. **I**

SVERDRUP, H. U., JOHNSON, M. W. and FLEMING, R. H. 1942. The oceans - their physics, chemistry, and general biology: 1077 pp. New York

THORSON, G. 1946. Reproduction and larval development of Danish marine bottom invertebrates: Medd. Komm. Danm. Fisk., Havunders¢g, ser: Plankton. **4**, No. 1

THORSON, G. 1950. Reproductive and larval ecology of marine bottom invertebrates: Biol. Rev. **25**, 1-45

VAN WAGENEN, R. G. and O'ROURKE, N. W. 1960. The self-propelled research vehicle: a second progress report: Univ. Washington Apply. Phys. Lab. **60-63**, 20 pp.

VERWEY, J. 1954. De mosel en zijn eisen (with summary in English): 13 pp. Overdruk Uit; Faraday. Groningen

WELLS, A. L. 1938. Some notes on the plankton of the Thames estuary: J. Anim. Ecol. **7**, 105-24

WHEATLAND, A. B. 1959. Some aspects of the carbon, nitrogen and sulphur cycles in the Thames estuary. 33-50. in. The effects of pollution on living material: Symposia of the Institute of Biology No. 8, ed. Yapp, W. B.

WIBORG, K. F. 1951. The whirling vessel, an apparatus for the fractionating of plankton samples: Rep. Norweg. Fish. Invest. **II**, 1-16

WIBORG, K. F. 1961. Estimations of number in the laboratory. Symposium on zooplankton production: Copenhagen, 1961, ed. J. H. F. and J. Corlett, Rapp. Cons. Explor. Mer. **153,** 232 pp.

WICKSTEAD, J. H. 1961. A quantitative and qualitative study of some Indo-West-Pacific plankton: Fishery publications No. 16.200 pp. Colonial Office, London

WILSON, D. P. 1952. The influence of the nature of the substratum on the metamorphosis of the larvae of marine animals, especially the larvae of *Ophelia bicornis*, Savigny: Ann. Inst. Oceanograph. **27**, 49-156

WINSOR, C. P. and CLARKE, G. L. 1940. A statistical study of variations in the catch of plankton nets: J. mar. Res. **3**, 1-34

YENTSCH, C. S., GRICE, G. D. and HART, A. D. 1961. Some opening-closing devices for plankton nets operated by pressure, electrical and mechanical action. Symposium on Zooplankton production: Copenhagen, 1961, ed. J. H. F. and J. Corlett, Rapp. Cons. Explor. Mer. **153**, 232 pp.

ZEUTHEN, E. 1947. Body size and metabolic rate in the animal kingdom with special regard to the marine microfauna: C. R. Lab. Carlsburg. sér. chim. **26,** 17-161

ZOBELL, C. E. 1941. Apparatus for collecting water samples from different depths for bacteriological analysis: J. mar. Res. **4**, 173

ZOBELL, C. E. 1946. Marine microbiology, 240 pp. Waltham, Mass.

References to Groups

COMPREHENSIVE WORKS

Fiches d'identification. ed. Jespersen, P. and Russell, F. S. *et al*, 1939 - Cons. int. Explor. Mer. Copenhagen

MASSUTI, M. and MARGALEF, R. 1950. Introduction al estudio del Plankton marino:182pp. Barcelona

Nordisches Plankton: Zoologischer Teil. Kiel and Leipzig, 1901

TRÉGOUBOFF, G. and ROSE, M. 1957. Manuel de planctonologie Méditerrannéene: Vol. I, 587pp. Vol. II (mainly figs.). Paris

PHYTOPLANKTON

HORNER, R. A. 2002 Taxonomic Guide to Some Common Marine Phytoplankton Biopress. 195pp.

HUSTEDT, F. 1930. Kieselagen - in Kryptogammen - Flora ed by Rabenhorst. **7**.

LEBOUR, M. V. 1930. The planktonic diatoms of the northern seas: Ray. Soc. 244pp. London.

PARKE, M. 1949. Studies on marine flagellates: J. mar. biol. Ass. U.K. **28**, 255-85

PARKE, M. 1961. Some remarks concerning the class Chrysophyceae: Brit. phycol. Bull. **2**, 47-57

Identifying Marine Phytoplankton. ed. Tomas, C.R. 1997, Academic Press. 858pp.

Marine Phytoplankton A Guide to Naked Flagellates and Coccolithophorids. ed. Tomas, C.R. , Throndsen, J. and Heimdal, B. 1993 327pp.

ZOOPLANKTON

Protozoans

BÉ, A. W. H. 1967. Foraminifera. Fiches d'identification du zooplancton. Sheet 108: Cons. int. explor. Mer. Zooplankton.

BOCK, K. J. 1967. Protozoa. Fiches d'identification du zooplancton. Sheet 110: Cons. int. plor. Mer. Zooplankton.

BOTTAZZI MASSERA, E and NENCINI, G. 1969. Acantharia. Fiches d'identification du zooplancton. Sheet 114: Cons. int. explor. Mer. Zooplankton.

CAMPBELL, A. S. 1942. The oceanic Tintinnoinea of the plankton gathered during the last cruise of the Carnegie. Pubis. Carnegie Inst. No. 537.

CHATTON, E. 1952. Classe des Dinoflagellés ou Peridiniens: Traité de Zoologie. Ed. Grassé, P. **1**.

CUSHMAN, J. A. 1949. Foraminifera. Their classification and economic use. Harvard. 605pp.

ICES Zooplankton Methodology Manual. ed. Harris, R., Skjoldal, H, Wiebe, P., Huntley, M. 2000 Academic Press 684pp.

KOFOID, C. A. and CAMPBELL, A. S. 1929. A conspectus of the marine and freshwater Ciliata belonging to the suborder Tintinnoinea with descriptions of new species principally from the Agassiz Expedition to the eastern tropical Pacific 1904-1905. Univ. Calif. Publ. Zool. **34**, 1-403.

KOFOID, C. A. and CAMPBELL, A. S. 1939. The Ciliata: The Tintinnoinea. Bull. Mus. comp. Zool. Narv. **84**, 1-473.

LEBOUR, M. V. 1925. The dinoflagellates of northern seas. Mar. biol. Ass. U.K. Plymouth. 250pp.

LE CALVEZ, J. 1953. Ordre de Foraminifères. Traité de Zoologie. Ed. Grasse, P. **1**.

MARSHALL, S. M. 1969. Protozoa: Family Tintinnidiidae. Fiches d'identification du zooplancton. Sheets 117-127: Cons. int. explor. Mar. Zooplankton.

TODD, C.D., LAVERACK, M.S., BOXSHALL, G. 1996 Coastal Marine Zooplankton *A Practical Manual for Students, Edition 2* Cambridge University Press 106pp.

TREGOUBOFF, G. 1953. Classe des Radiolaires. Traité de Zoologie. Ed. Grassé, P. **1**.

TREGOUBOFF, G. 1953. Classe des Heliozoa. Traité de Zoologie. Ed. Grassé, P. **1**.

VAN COUWELAAR, M. 2003 Zooplankton and Micronekton of the North Sea *Version 1* Series: WORLD BIODIVERSITY DATABASE Illus. ETI Information Services Ltd. CD (PC/Mac hybrid)

Coelenterates

Medusae

KRAMP, P. L. 1961. Synopsis of the medusae of the world: J. mar. biol. Ass. U.K. **40**, 468pp.

MAYER, A. G. 1910. Medusae of the world: 3 vols. Carnegie Inst., Washington

RUSSELL, F. S. 1939-55. Hydromedusae - Fiches d'identification du zooplancton. Several sheets: Cons. int. Explor. Mer.

RUSSELL, F. S. 1953. The medusae of the British Isles: 530pp. Cambridge

RUSSELL, F. S. 1970. Hydromedusae. Fiches d'identification du zooplancton. Sheet 128: Cons. int. Explor. Mer.

Siphonophores

GARSTANG, W. 1946. The morphology and relations of the Siphonophora: Quart. J. micr. Sci. **87**, 103-93

HYMAN, L. H. 1940. Protozoa through Ctenophora: 726pp. New York

KIRKPATRICK, P.A., PUGH, P.R. 1984 SBF Volume 29: Siphonophores and Velellids Series: SYNOPSES OF THE BRITISH FAUNA Field Studies Council 154pp.

TOTTON, A. K. and FRASER, J. H. 1955. Siphonophora - Fiches d'identification du Zooplancton. Several sheets: Cons. int. Explor. Mer.

Scyphozoans

MAYER, A. G. 1910. Medusae of the world Vol. 3: Scyphomedusae. Carnegie Inst., Washington

Ctenophores

LILEY, R. 1958. Ctenophora - Fiches d'identification du zooplancton. Sheet 82.: Cons. int. Explor. Mer.

VANHÖFFEN, E. 1903. Ctenophoren: Nord. Plankt. **11**, 1-7

Polychaetes

CHAMBERS, S. J., MUIR, A.I. 1997 SBF Volume 54: Polychaetes: British Chrysopetaloidea, Pisionoidea and Aphroditoidea Series: SYNOPSES OF THE BRITISH FAUNA Field Studies Council. 202pp.

FAUVEL, P. 1923. Polychètes errantes: Faune Fr. **5**, Paris

GEORGE, J. D., HARTMANN-SCHRODER, G. 1985 SBF Volume 32: Polychaetes British Amphinomida, Spintherida and Eunicida Series: SYNOPSES OF THE BRITISH FAUNA Field Studies Council. 221pp.

HAMMOND, R. 1967. Polychaeta: Syllidae. Fiches d'identification du zooplancton. Sheet 113: Cons. int. Explor. Mer.

MUUS, B. J. 1953. Polychaeta - Fiches d'identification du zooplancton. Sheets 52 and 53: Cons. int. Explor. Mer.

ROUSE, G., PLEIJEL, F. 2001 Polychaetes Oxford University Press. 354pp.

WESENBERG-LUND, E. 1935. Tomopteridae and Typhloscolecidae: Danish Ingolf-Exped. **4**, No. II

WESENBERG-LUND, E. 1939. Pelagic polychaetes of the families Aphroditidae, Phyllodicidae, Typhloscolecidae and Alciopidae: Rep. Danish. Oceanogr. Exped. Medit. 1908-10 **2**, No. 2

Polychaetes *An Interactive Identification Guide* ed. Wilson, R. S., Hutchings, P. A. 2003 CSIRO CD

Chaetognaths

RUSSELL, F. S. 1939. Chaetognatha - Fiches d'identification du zooplancton. Sheet **I**: Cons. int. Explor. Mer.

FRASER, J. H. 1957. Revision of above sheet

Crustacea (general)

CALMAN, W. T. 1909. Crustacea *in* A treatise on zoology: ed. E. Ray Lancaster. Oxford

CALMAN, W. T. 1911. The life of Crustacea: 289pp. London.

Cladocerans

GURNEY, R. 1927. Copepoda and Cladocera of the Plankton: Trans. zool. Soc. Lond. part 2

RAMMNER, W. 1939. Cladocera - Fiches d'identification du zooplancton. Sheet 3: Cons. int. Explor. Mer.

Ostracods

ANGEL, M. V. 1993 SBF Volume 48: Marine Planktonic Ostracods Series: SYNOPSES OF THE BRITISH FAUNA Field Studies Council. 240pp.

ANGEL, M. V. 2000 Marine Planktonic Ostracods Mac/Windows Version Series: WORLD BIODIVERSITY DATABASE ETI Information Services Ltd. CD (PC/Mac hybrid)

KLIE, W. 1944. Ostracoda - Fiches d'identification du zooplancton. Sheets 6 and 7: Cons. int. Explor. Mer.

POULSEN, E. M. 1969. Ostracoda - Fiches d'identification du zooplancton. Sheets 115 and 116: Cons. int. Explor. Mer.

SARS, G. O. 2005 Account of the Crustacea of Norway, Vol. IX: Ostracoda. Pisces Conservation Ltd. CD

Copepods

BOXSHALL, G. A., HALSEY, S. H. 2004 An Introduction to Copepod Diversity. 2 vols Ray Society. 970pp.

FARRAN, G. P. 1948. Fiches d'identification du zooplancton. Several sheets: Cons. int. Explor. Mer.

HARDING, J. P., SMITH, W. A. 1974 A Key to the British Freshwater Cyclopid and Calanoid Copepods Series: FRESHWATER BIOLOGICAL ASSOCIATION SCIENTIFIC PUBLICATIONS 18 Freshwater Biological Association 56pp.

HUYS, R., GEE, J. M., MOORE, C. G. and HAMOND, R. 1996 SBF Volume 51: Marine and Brackish Water Harpacticoid Copepods, Part 1 Series: SYNOPSES OF THE BRITISH FAUNA Field Studies Council. 352pp.

KLIE, W. 1943. Copepoda Harpacticoida - Fiches d'identification du zooplancton. Several sheets: Cons. int. Explor. Mer.

ROSE, M. 1933. Copépodes pélagiques: Faune Fr. **26**, Paris

VERVOORT, W. 1952. Fiches d'identification du zooplancton. Several sheets: Cons. int. Explor. Mer.

WELLS, J. B. J. 1970. Copepoda - Harpacticoida. Fiches d'identification du zooplancton. Sheet 133: Cons. int. Explor. Mer.

Isopods

NAYLOR, E. 1957. Isopoda - Fiches d'identification du zooplancton. Sheets 77 and 78: Cons. int. Explor. Mer.

NAYLOR, E. 1972 SBF Volume 3: British Marine Isopods Series: SYNOPSES OF THE BRITISH FAUNA Linnean Society of London. 90pp.

Cumaceans

JONES, N. S. 1957. Cumacea - Fiches d'identification du zooplancton. Sheets 71-76: Cons. int. Explor. Mer.

Amphipods

BOWMAN, T. E. 1960. The pelagic amphipod genus *Parathemisto* (Hyperiidea, Hyperiidae) in the North Pacific and adjacent Arctic Ocean: Proc. U.S. nat. Mus. **112**, 343-92

CHEVREUX, E. and FAGE, L. 1925. Amphipodes: Faune Fr. **9**, Paris. 488pp.

CHANG-TAI SHIM and DUNBAR, M. J. 1963. Amphipoda: Phronimidae. Fiches d'identification du zooplancton. Sheet 104: Cons. int. Explor. Mer.

DUNBAR, M. J. 1963. Amphipoda: Sub-order Myperiidea. Fiches d'identification du zooplancton. Sheet 103. Cons. int. Explor. Mer.

SARS, G. O. 2006 Account of the Crustacea of Norway, Vol. I: Amphipoda. Pisces Conservation Ltd. CD

SCHELLENBERG A. 1927. Amphipoda des Nordischen Plankton: Nord. Plankt. **6**, 589-722

Mysids

NOUVEL, H. 1950. Mysidacea - Fiches d'identification du zooplancton. Sheets 18-27: Cons. int. Explor. Mer.

TATTERSALL, W. M. and TATTERSALL, O. S. 1951. The British Mysidacea. 460pp.: Ray. Soc. London

TATTERSALL, W. M., and TATTERSALL, O. S. 2003 British Mysidacea. Pisces Conservation Ltd. CD

Euphausids

EINARSSON, H. 1945. Euphausiacea - North Atlantic species: Dana. Rep **5**, No. 27. 185pp.

GLOVER, R. S. 1952. Continuous plankton records: The Euphausiacea of the North-Eastern Atlantic and the North Sea, 1946-48: Bull. Mar. Ecol. **3**, No. 23, 185-214

HOLT, E. W. L. and TATTERSALL, M. M. 1905. Schizopodous Crustacea from the N.E. Atlantic slope: Fish. Sci. Invest., Ireland. **4**, No. 1

VON ZIMMER, C. 1932. Euphausiacea *in* Die Tierwelt der Nord- and Ostsee. **10**, 10-28

Decapods

RICE, A. L. 1967. Crustacea (Pelagic Adults). Order Decapoda. Fiches d'identification du zooplancton. Sheet 112: Cons. int. Explor. Mer.

Molluscs

Pteropods

MORTON, J. E. 1957. Gymnosomata - Fiches d'identification du zooplancton. Sheets 79 and 80: Cons. int. Explor. Mer.

RAMMNER, W. 1939. Cladocera - Fiches d'identification du zooplancton. Sheet 3: Cons. int. Explor. Mer.

Ostracods

KLIE, W. 1944. Ostracoda - Fiches d'identification du zooplancton. Sheets 6 and 7: Cons.

BÜCKMANN, A. 1969. Appendicularia - Fiches d'identification du zooplancton. Sheet 7 (Revised): Cons. int. Explor. Mer.

BÜCKMANN, A. 1945. Appendicularia - Fiches d'identification du zooplancton. Sheet 7: Cons. int. Explor. Mer.

FRASER, J. H. 1947. Thaliacea - Fiches d'identification du zooplancton. Sheets 9 and 10: Cons. int. Explor. Mer.

INVERTEBRATE LARVAE

Polychaete larvae

ANDERSON, D. T. 1961. The development of the polychaete *Haploscoloplos fragilis*: Quart. J. micr. Sci. **102**, 257-72

GRAVELY, F. H. 1909. Polychaete larvae: L.M.B.C. Memoir **19**

HANNERZ, L. 1956. Larval development of the polychaete families, Spionidae Sars, Disomidae Mesnil and Poecilochaetidae N. fam. in the Gullmar Fjord (Sweden): Zool. Bidr. Uppsala. **31**, 1-204

HANNERZ, L. 1961. Polychaeta: Larvae. Families Spionidae, Disomidae, Poecilochaetidae - Fiches d'identification du zooplankton. Sheet 91: Cons. int. Explor. Mer.

NOLTE, W. 1936 and 1938. Annelidenlarven: Nord. Plankt. **5**, 23, 59-282

RASMUSSEN, E. 1956. Faunistic and biological notes III. The reproduction and larval development of some polychaetes from Isefjord, with some faunistic notes: Biol. Medd. Dan. Vid. Selsk. **23**, 1-84

SHEARER, C. 1911. On the development and structure of the trochophore of *Hydroides uncinatus* (*Eupomatus*): Quart. J. micr. Sci. **56**, 543-90

WILSON, D. P. 1932. On the mitraria larva of *Owenia fusiformis* Delle Chiaje. Philos. Trans. B. **221**, 231-4

Crustacea

Cirripedes

BASSINDALE, R. 1936. The developmental stages of three English barnacles - *Balanus balanoides* (Linn.), *Chthamalus stellatus* (Poli), and *Verruca stroemia* (O. F. Müller): Proc. zool. Soc. Lond. 57-74

KNIGHT-JONES, E. W. and WAUGH, G. D. 1949. On the larval development of *Elminius modestus* Darwin: J. mar. biol. Ass. U.K. **28**, 413-28

Euphausids

LEBOUR, M. V. 1926. A general survey of larval Euphausids with a scheme for their identification: J. mar. biol. Ass. U.K. **14**, 519-23

General and Shrimps and Prawns

The Recent Genera of the Caridean and Stenopodidean Shrimps Crustacea, Decapoda. With an appendix on the order Amphionidacea ed. Fransen, C.H.J.M., van Achterberg, C., Holthuis., L.B. 1993 Series: ZOOLOGISCHE VERHANDELINGEN 26 Nationaal Natuurhistorisch Museum. 328pp.

GURNEY, R. 1903*a*. The metamorphosis of the Decapod Crustaceans *Aegeon* (*Crangon*) *fasciatus* Risso, and *Aegeon* (*Crangon*) *trispinosus*, Hailstone: Proc. zool. Soc. Lond. 24-30

GURNEY, R. 1903*b*. The larvae of certain British Crangonidae: J. mar. biol. Ass. U.K. **6**, 595-7

GURNEY, R. 1923*a*. Some notes on *Leander longirostris* M. Edwards, and other British prawns: Proc. zool. Soc. Lond. 97-123

GURNEY, R. 19236. The larval stages of *Processa canaliculata*, Leach.: J. mar. biol. Ass. U.K. **12**, 245-65

GURNEY, R. 1924*a*. The larval development of some British prawns (Palaemonidae). I *Palaemonetes varians*: Proc. zool. Soc. 297-328

GURNEY, R. 1924*b*. The larval development of some British prawns (Palaemonidae). II. *Leander longirostris* and *L. squilla*: Proc. zool. Soc. Lond. 961-82

GURNEY, R. 1926. The protozoeal stage in Decapod development: Ann. Mag. nat. Hist. (9) **18**, 19-27

GURNEY, R. 1939. Bibliography of the larvae of Decapod Crustacea: Ray. Society Lond. **125**, 306 pp.

LEBOUR, M. V. 1931. The larvae of the Plymouth Caridea. I. The larvae of the Crangonidae. II. The larvae of the Hippolytidae: Proc. zool. Soc. Lond. 1-9

LEBOUR, M. V. 1936. Notes on the Plymouth *Processa*: Proc. zool. Soc. Lond. 609-17

LEBOUR, M. V. 1940. The larvae of the Pandalidae: J. mar. biol. Ass. U.K. **24**, 239-52

MACDONALD, J. D., PIKE, R. B. and WILLIAMSON, D. I. 1957. Larvae of the British species of *Diogenes*, *Pagurus*, *Anapagurus* and *Lithodes* (Crustacea, Decapoda): Proc. zool. Soc. Lond. **128**, 209-57

PIKE, R. B. and WILLIAMSON, D. I. 1958. Crustacea Decapoda: Larvae. Fiches d'identification du zooplancton. Sheet 81: Cons. int. Explor. Mer.

SBF Volume 15: Coastal Shrimps and Prawns Keys and Notes for Identification of the Species ed. Smaldon, G., Holthius, L. B., Fransen, C. H. J. M. 1994 Series: SYNOPSES OF THE BRITISH FAUNA . Field Studies Council *Edition 2.* 142pp.

WILLIAMSON, D. I. 1957. Crustacea Decapoda: Larvae. Fiches d'identification du zooplancton. Sheets 67 and 68: Cons. int. Explor. Mer.

WILLIAMSON, D. I. 1960. Crustacea Decapoda: Larvae. Fiches d'identification du zooplancton. Sheet 90: Cons. int. Explor. Mer.

WILLIAMSON, D. I. 1962. Crustacea Decapoda: Larvae. Fiches d'identification du zooplancton. Sheet 92: Cons. int. Explor. Mer.

WILLIAMSON, D. I. 1967. Crustacea Decapoda: Larvae. Fiches d'identification du zooplancton. Sheet 109: Cons. int. Explor. Mer.

Anomura

LEBOUR, M. V. 1930. The larvae of the Plymouth Galatheidae I: J. mar. biol. Ass. U.K. **17**, 175-86

LEBOUR, M. V. 1931. The larvae of the Plymouth Galatheidae II: J. mar. biol. Ass. U.K. **17**, 385-90

WEBB, G. E. 1919. The development of the species of *Upogebia* from Plymouth Sound: J. mar. biol. Ass. U.K. **12**, 81-134

WEBB, G. E. 1921. The larvae of the Decapoda, Macrura and Anomura of Plymouth: J. mar. biol. Ass. U.K. 385-413

Brachyura

LEBOUR, M. V. 1927. Studies of the Plymouth Brachyura I: J. mar. biol. Ass. U.K. **14**, 795-821

LEBOUR, M. V. 1928*a*. The larval stages of the Plymouth Brachyura: Proc. zool. Soc. Lond. 473-560

LEBOUR, M. V. 1928*b*. Studies of the Plymouth Brachyura II: J. mar. biol. Ass. U.K. **15**, 109-18

Molluscs

FRETTER, V. and PILKINGTON, M. C. 1970. Prosobranchia: veliger larvae of Taenioglossa and Stenoglossa. Fiches d'identification du zooplancton. Sheets 129-132: Cons. int. Explor. Mer.

HADFIELD, M. G. 1964. Opisthobranchia. Fiches d'identification du zooplancton. Sheet 106: Cons. int. Explor. Mer.

Papers by LEBOUR, M. V. 1931-8 in J. mar. biol. Ass. U.K. and Proc. zool. Soc. Lond.

REES, C. B. 1950. The identification and classification of lamellibranch larvae: Bull. Mar. Ecol. **3**, No. 19, 73-104

Ectoprocta

RYLAND J. S. 1965 Polzoa (Bryozoa). Fiches d'identification de zoaplankton. Sheet 107: Cons. int. Explor. Mer.

Echinoderms

HORSTADIUS, Sv. 1939. Entwicklung von *Astropecten aranciacus*: Pubbl. Staz. zool. Napoli. **17**

MORTENSEN, TH. 1901. Die Echinodermen-larven: Nord. Plankt. **5**, 9, 1-30

MORTENSEN, Th. 1931, 1937, 1938. Contributions to the study of the development and larval forms of Echinoderms I–IV. Kgl. Danske. Vidensk. Selsk Skrifter, Naturv. Math. Afd. **4** and **7**

Phoronid larvae

FORNERIS, L. 1957. Family Phoronidae, Actinotrocha Larvae. Fiches d'identification du zooplancton. Sheet 69: Cons. int. Explor. Mer.

Hemichordate larvae

BURDON-JONES, C. 1957. Hemichordata, Family Ptychoderidae: Tornaria Larvae. Fiches d'identification du zooplancton. Sheet 70: Cons. int. Explor. Mer.

HYMAN, L. H. 1959. The Invertebrates: V. Smaller Coelomate Groups: McGraw Hill

Fish eggs and Larvae

EHRENBAUM, A. 1905-1909. Eier and Larven von Fischen. Nord. Plankt. 1, pp. 413

MCINTOSH, W. C. and MASTERMAN, A. T. 1897. The life-histories of the British Marine food fishes: pp. 467 London (C. J. Clay & Sons)

MUNK, P. and NIELSON, J. G. 2005 Eggs and larvae of North Sea fishes. Biofolia, Frederiksberg, Denmark. 224 pp.

SIMPSON, A. C. 1956. The pelagic phase: pp. 207-250 *in* Sea fisheries, ed. Graham, M. London

RUSSELL, F. S. 1973. A summary of the observations on the occurrence of planktonic stages of fish off Plymouth, 1924-1972. J. mar. biol. Ass. U.K. **53**, 347-355

RUSSELL, F. S. 1976. The Eggs and Planktonic stages of British Marine Fishes: pp. 524 London (Academic Press)

Index

A

B

C

H

I

J

K

L

M

R

S

T

U

V

X

Z

Made in United States
Orlando, FL
04 November 2026

90207700R00131